Machine Learning and Data Mining Annual Volume 2023

Edited by Marco Antonio Aceves-Fernández

Published in London, United Kingdom

Machine Learning and Data Mining Annual Volume 2023
http://dx.doi.org/10.5772/intechopen.113978
Edited by Marco Antonio Aceves-Fernández

Contributors
Bruno Menezes, Erick Sperandio, Júnia Ortiz, Ricardo de Oliveira, Michael D. Dong Wang, Janine Zitianellis, Hany Mohamed Nabil Helmy, Sherif El Diasty, Hazem Shatila, Luwei Li, Ali Akbar Firoozi, Ali Asghar Firoozi

First published in London, United Kingdom, 2023 by IntechOpen
IntechOpen is the global imprint of INTECHOPEN LIMITED, registered in England and Wales, registration number: 11086078, 5 Princes Gate Court, London, SW7 2QJ, United Kingdom

British Library Cataloguing-in-Publication Data
A catalogue record for this book is available from the British Library

Additional hard and PDF copies can be obtained from orders@intechopen.com

Machine Learning and Data Mining Annual Volume 2023
Edited by Marco Antonio Aceves-Fernández
p. cm.

This title is part of the Artificial Intelligence Book Series, Volume 21
Series Editor: Andries Engelbrecht

Print ISBN 978-0-85014-513-7
Online ISBN 978-0-85014-514-4
eBook (PDF) ISBN 978-0-85014-515-1
ISSN 2633-1403

For EU product safety concerns:
IN TECH d.o.o., Prolaz Marije Krucifikse Kozulić 3, 51000 Rijeka, Croatia,
info@intechopen.com or visit our website at intechopen.com.

IntechOpen Book Series
Artificial Intelligence
Volume 21

Aims and Scope of the Series

Artificial Intelligence (AI) is a rapidly developing multidisciplinary research area that aims to solve increasingly complex problems. In today's highly integrated world, AI promises to become a robust and powerful means for obtaining solutions to previously unsolvable problems. This Series is intended for researchers and students alike interested in this fascinating field and its many applications.

Meet the Series Editor

Andries Engelbrecht received the Masters and Ph.D. degrees in Computer Science from the University of Stellenbosch, South Africa, in 1994 and 1999 respectively. He is currently appointed as the Voigt Chair in Data Science in the Department of Industrial Engineering, with a joint appointment as Professor in the Computer Science Division, Stellenbosch University. Prior to his appointment at Stellenbosch University, he has been at the University of Pretoria, Department of Computer Science (1998-2018), where he was appointed as South Africa Research Chair in Artifical Intelligence (2007-2018), the head of the Department of Computer Science (2008-2017), and Director of the Institute for Big Data and Data Science (2017-2018). In addition to a number of research articles, he has written two books, Computational Intelligence: An Introduction and Fundamentals of Computational Swarm Intelligence.

Meet the Topic Editor

Dr. Marco Antonio Aceves-Fernández obtained his B.Sc. (Eng.) in Telematics from the Universidad de Colima, Mexico. He obtained both his M.Sc. and Ph.D. from the University of Liverpool, England, in the field of Intelligent Systems. He is a full professor at the Universidad Autonoma de Queretaro, Mexico, and a member of the National System of Researchers (SNI) since 2009. Dr. Aceves-Fernández has published more than 80 research papers as well as a number of book chapters and congress papers. He has contributed to more than 20 funded research projects, both academic and industrial, in the area of artificial intelligence, ranging from environmental, biomedical, automotive, aviation, consumer, and robotics applications. He is also an Honorary President of the National Association of Embedded Systems (AMESE), a member of the Mexican Academy of Computing (AMEXCOMP), a senior member of the IEEE, and a board member of many institutions and associations. His research interests include intelligent and embedded systems.

Contents

Preface **XIII**

Section 1
Machine Learning and Data Mining **1**

Chapter 1 **3**
Complexity Analysis in Channel Estimation Massive MIMO Compared with LMU and GRU
by Hany Helmy, Sherif El Diasty and Hazem Shatila

Section 2
Current Applications and Future Trends **13**

Chapter 2 **15**
Artificial Intelligence Techniques for Political Risk Management: An NLP Analysis of the 2019 US-China Trade War
by Michael D. Wang

Chapter 3 **45**
A Quantitative Analysis of Big Data Analytics Capabilities and Supply Chain Management
by Janine Zitianellis

Chapter 4 **75**
Application of Machine Learning in Geotechnical Engineering for Risk Assessment
by Ali Akbar Firoozi and Ali Asghar Firoozi

Chapter 5 **105**
Application of Machine Learning and Data Mining in Medicine: Opportunities and Considerations
by Luwei Li

Chapter 6 **123**
Automatic BI-RADS Classification of Breast Magnetic Resonance Medical Records Using Transformer-Based Models for Brazilian Portuguese
by Ricardo de Oliveira, Bruno Menezes, Júnia Ortiz and Erick Nascimento

Preface

Implementing AI applications can be highly complex due to various factors. Firstly, acquiring and preprocessing large volumes of high-quality data necessary for training AI models is a significant challenge. Additionally, selecting the appropriate algorithms, architectures, and frameworks demands expertise as each application may require a different approach. Balancing computational resources, such as processing power and memory, is crucial for efficient model training and deployment. Moreover, ensuring ethical considerations, addressing bias, and maintaining transparency and interpretability in AI systems add another layer of complexity.

The ability of machines to demonstrate advanced cognitive skills in making decisions, learning, perceiving the environment, predicting certain behaviors, or processing written or spoken languages, among other skills, makes this discipline of paramount importance in today's world.

Deploying AI applications poses several challenges. One significant hurdle is understanding how AI systems arrive at decisions or predictions. Especially in critical domains like healthcare or finance, this understanding remains a challenge. Addressing these challenges requires a multidisciplinary approach involving expertise in data science, domain knowledge, ethics, and ongoing regulatory adaptation.

This book compiles a range of applications in Machine Learning in two sections: 'Machine Learning and Data Mining' and 'Current Applications and Future Trends'.

I hope that this work will be of interest to students and researchers alike, as I did my best to compose quality research contributions with a number of different applications.

Marco Antonio Aceves-Fernández, Ph.D.
Faculty of Engineering,
Universidad Autónoma de Querétaro,
Querétaro, México

Section 1

Machine Learning and Data Mining

Chapter 1

Complexity Analysis in Channel Estimation Massive MIMO Compared with LMU and GRU

Hany Helmy, Sherif El Diasty and Hazem Shatila

Abstract

MIMO: Multiple-input multiple-output technology uses multiple antennas to use reflected signals to provide channel robustness and throughput gains. It is advantageous in several applications like cellular systems, users are distributed over a wide coverage area in various applications such as mobile systems, improving channel state information (CSI) processing efficiency in massive MIMO systems. This chapter proposes two channel-based deep learning methods gated recurrent unit and a Legendre memory unit to enhance the performance in a massive MIMO system and compares the complexity analysis to the previous methods, The complexity analysis is based on the channel state information network combined with gated recurrent units and Legendre memory units compared to indicator parameters which show the difference between literature-based techniques.

Keywords: massive MIMO, FDD, compressed sensing, deep learning, conventional neural network, complexity analysis, gated recurrent unit, Legendre memory unit

1. Introduction

Complexity analysis in deep learning is a fascinating topic indeed when delving into the realm of deep learning, we often encounter intricate models with numerous layers and countless parameters [1]. It is essential for such models to truly comprehend their inner workings. Now, complexity analysis in deep learning involves assessing the computational requirements and efficiency of these models. One commonly used metric is time complexity, which examines how the computational costs increase with an increase in the input size [2]. It allows us to estimate the computational resources needed for training and inference tasks.

Another aspect to consider is the space complexity, which examines the memory requirements of deep learning models. As the number of layers and parameters grows, so does the memory needed to store them [3]. Understanding this helps us optimize memory usage and select suitable hardware for our neural networks. However, it's worth mentioning that deep learning is an ever-evolving field, and complexity analysis is just a small part of the puzzle.

IntechOpen

There are always new challenges, surprises, and complexities waiting to be unraveled. It's a journey that continues to intrigue and captivate scientists and enthusiasts alike. The history of complexity analysis in deep learning. It is indeed an intriguing subject. The path to understanding the complexity of deep learning has been paved with both triumphs and challenges. Back in the early days, when I first ventured into the realm of theoretical physics, we were only scratching the surface of what would later become known as deep learning [4]. The concept of neural networks has been around for quite some time, but it was the advent of powerful, computational capabilities that truly allowed us to explore their potential. As we delved deeper, we began to realize that the complexity of deep learning models was no trivial matter.

The number of parameters, the intricate interconnections, and the sheer depth of these networks made it a fascinating puzzle to unravel. The primary challenge was to develop methods for analyzing and understanding the complexity of these models. We needed tools that could help us comprehend the behavior of deep learning systems, predict their performance, and discern their limitations. Over time, researchers developed various approaches to complexity analysis in deep learning [5]. They ranged from straightforward measures like counting the number of parameters or layers, to more sophisticated techniques such as analyzing the network's computational and memory requirements. Additionally, advancements were made in characterizing the computational complexity of training deep learning models. The notion of time and resource complexity became crucial in comprehending the training process and predicting how long it might take, given a certain dataset and architecture [6].

2. Complexity analysis

2.1 Performance and complexity trade-off of LMU and GRU

The Legendre Memory Unit (LMU) is mathematically derived to orthogonalize its continuous-time history – doing so by solving d coupled ordinary differential equations (ODEs), whose phase space linearly maps onto sliding windows of time via the Legendre polynomials up to degree $d-1$.

The Gated Recurrent Unit (GRU) is a type of Recurrent Neural Network (RNN) that, in certain cases, has advantages over long short-term memory (LSTM) (**Figure 1**). GRU uses less memory and is faster than LSTM, however, LSTM is more accurate when using datasets with longer sequences (**Figure 2**).

The RNN layers for CSI compression and reconstruction of massive MIMO systems might have a remarkable number of parameters. The recurrent neural network module, for example, can dd up to $\mathbf{10^8}$ more parameters, raising storage and computation difficulties. To reduce the computational complexity and required memory, a fully connected layer-based autoencoder has been developed for channel state information feedback in time-varying channels [9].

While other research has investigated smaller recurrent neural networks RNNs, the least computationally expensive of these models still needs $\mathbf{10^7}$ parameters per snapshot. When the compression ratio is small, the networks also suffer from a significant drop in feedback performance because they must keep the same compression ratio in succeeding time slots and cannot obtain accurate previous information in the first time slot [10]. Despite the reported performance of "stacked" GRUs, the

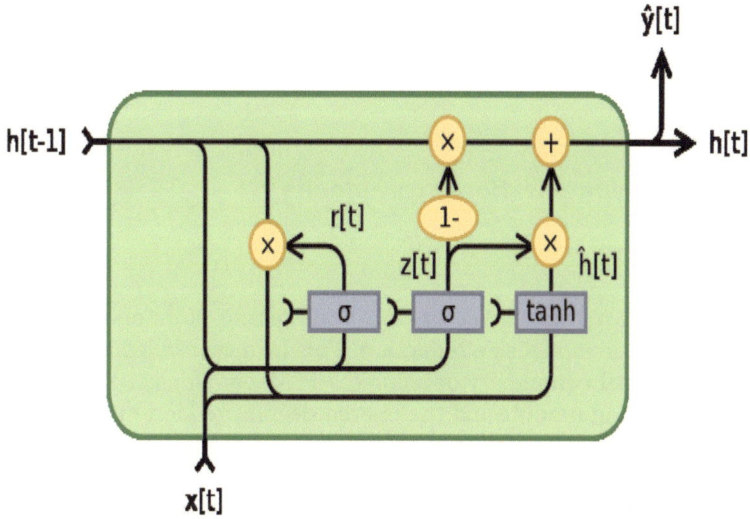

Figure 1.
The block diagram of (GRU) [7].

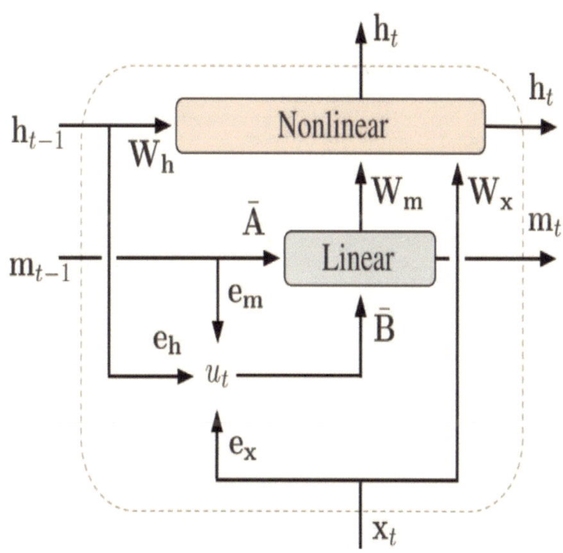

Figure 2.
Time-unrolled LMU layer [8].

minimum depth of recurrent layers required for CSI recovery accuracy has yet to be determined [11]. Despite the substantial number of RNN parameters, the performance gain should be significant enough to justify the memory overhead.

The practical constraint of how often such feedback can be transmitted, as well as how CSI of fading channels will vary due to the Doppler effect, should be considered in DNN-based CSI feedback and recovery [12]. To reduce computational complexity and model size, we seek to systematically exploit channel characteristics such as forward CSI temporal coherence. We directly leverage known channel coherence temporally by developing a simple but effective Legendre memory unit combined with gated recurrent unit channel state information feedback LMU to improve CSI recovery accuracy.

3. Model size and computational complexity

3.1 The structure

Table 1 shows the number of parameters and floating-point operations (FLOPs) to show the impact of the model size reduction. This information compares the storage size and computational complexity of using the FC-layer along with the CNN-layer in the CsiNet compression module and the related decompression module.

The CNN-layer-based dimension compression and decompression module reduces the number of parameters by more than 100 times and the number of FLOPs by at least four times. The comparison results show that our CNN architecture for CSI compression and decompression is a significant step forward in expanding the spectrum of practical applications for deep learning-based channel state information encoding, feedback, and reconstruction in massive MIMO wireless systems.

Despite the higher cost, CsiNet-GRU and LMU-GRUs can reduce the number of parameters by order of magnitude and save over (5/6) FLOPs over CsiNet. **Table 2**

	Number of parameters		FLOPs	
	FC-based	**CNN-layer**	**FC-based**	**CNN-layer**
CR = 1/4	2.1 M	14.4 K	4.2 M	0.9 M
CR = 1/16	1.1 M	7.2 K	2.1 M	0.5 M
CR = 1/32	0.5 M	3.7 K	1.0 M	0.2 M
CR = 1/64	0.2 M	2.5 K	0.5 M	0.1 M

M, million; K, thousand.

Table 1.
Number of parameters and FLOPs comparison for FC-based and CNN-based dimension compression and decompression module.

	γ	CsiNet [12]	Conv-LSTM-CsiNet [13]	CsiNet-GRU [7]	LMU-GRUs
Params	1/4	2,103,904	28,326,904	2,107,684	2,108,992
	1/16	1,055,072	22,296,312	1,058,852	1,060,160
	1/32	530,656	19,477,624	534,436	535,744
	1/64	268,448	18,117,432	272,228	273,536
MACCs	1/4	4,366,336	121,708,544	12,424,448	12,497,120
	1/16	3,842,048	97,591,296	10,375,872	11,448,544
	1/32	3,579,904	86,319,104	9,851,548	10,924,256
	1/64	3,448,832	80,879,616	9,589,440	10,662,112

Table 2.
The number of parameters and MACCS.

shows the complexity analysis of the proposed LMU-GRUs in comparison to various state-of-the-art CSI feedback systems, where the number of parameters and MACCs1 stand for space and temporal complexity, respectively [14]. When compared to CsiNet, our frameworks add fewer parameters while considerably improving CSI recovery quality, as seen in **Table 2**.

Convolution layers are mainly responsible for the increase in MACCs. When CR is low, convolutional layers require more computation than dense layers. LMU-GRUs have much lower model parameters and MACCs than CsiNet and Conv-LSTM-CsiNet, which improve reconstruction accuracy at the expense of huge space and temporal complexity due to the dense layers in GRU cells [15].

Each network's average number of floating-point operations (FLOPs) connected with a single timeslot is also shown in **Table 3**. When compared to the Conv-LSTM-CsiNet in each compression ratio, CsiNet-GRU, and LMU-GRUs can save more than (8/9) and 9/10) FLOPs, respectively. Even at low compression ratios, the amount of computation for Conv-LSTM-CsiNet does not decrease considerably [16].

Without loss of performance, we show that latent convolutional layers require much fewer parameters than FC layers. **Table 3** compares the model size and computational complexity of CsiNet, Conv-LSTM-CsiNet, CsiNet-GRU, and LMU-GRUs of a single timeslot, respectively.

LMU-GRUs utilize 60 times fewer parameters than Conv-LSTM-CsiNet across all compression ratios. More crucially, LMU-GRUs can achieve greater CSI recovery accuracy while using 1/3000 of the number of parameters required by Conv-LSTM-CsiNet. A 16-fold reduction in compression ratio (from 1/4 to 1/64), for example, only saves 1% of FLOPs. CsiNet-GRU and CsiNet-LMU-GRUs, on the other hand, require significantly less computing complexity in proportion to lower compression ratios.

A 16-fold CR reduction (from 1/4 to 1/64), for example, reduces the number of FLOPs in CsiNet-GRU and LMU-GRUs networks by 9 and 2%, respectively.

When using CsiNet-GRU and LMU-GRUs as a cooperative learning method at UE, we found that 50 percent more parameters and FLOPs are required than during the training phase. This is because the trained decoder must be repeated on the UE side to generate decoded CSI for the encoder's previous time slot.

	γ	CsiNet [12]	Conv-LSTM CsiNet [13]	CsiNet-GRU [7]	LMU-GRUs
Params	1/4	2.1 M	2.1 M	132.7 M	2.1 M
	1/16	1.0 M	542.9 K	118.5 M	1.0 M
	1/32	530.6 K	280.7 K	116.1 M	535.7 K
	1/64	268.4 K	149.6 K	115.0 M	273.5 K
FLOPs	1/4	412.9 M	44.5 M	41.2 M	39.2 M
	1/16	409.8 M	41.3 M	40.5 M	39.3 M
	1/32	409.2 M	40.8 M	40.4 M	40.1 M
	1/64	409.0 M	40.5 M	40.3 M	40.0 M

M, million; K, thousand.

Table 3.
Model size and computational complexity of tested networks.

Table 3 summarizes the number of parameters and floating-point operations (FLOPs) to show the impact of the proposed model size reduction. This data compares the storage space and computational difficulty of using the FC layer against the suggested CNN layer in the CSI compression module and the related decompression module.

The proposed LMU-GRUs dimension compression and decompression module, as shown in **Table 3**, reduces the number of parameters by over 100 times and the number of FLOPs by at least 4 times [17]. The comparison results show that our new LMU-GRUs architecture for CSI compression and decompression is a significant step forward in expanding the range of practical applications for deep learning-based CSI encoding, feedback, and reconstruction in massive MIMO wireless systems.

4. Conclusion

This chapter compares a channel state information (CSI) feedback network by extending the DL-based CsiNet technique to incorporate GRUs and LMU over GRU layers. The proposed LMU-GRUs technique achieved the number of parameters and MACCS compared to other CS-based and CSI-based methods. The work is motivated by the active use of the recurrent convolutional neural network (RCNN) model in model size and computational complexity of tested networks using a channel state information network. The basic concept is to compare the complexity analysis in deep learning techniques using the COST 2100 model to get results related to indicators parameters in time-varying MIMO channels and acquire training samples.

5. Recommendations for future work

Quantum-assisted Deep Learning (QDL) is receiving significant attention towards enhancing various performance metrics of communication networks. The classical DL faces various challenges; where a substantial challenge is to figure out the training method for complex topologies of neural networks (NNs) (which are of similar complexity to that of the natural structure of the human brain).

Acknowledgements

First of all, I thank ALLAH for giving me the will to achieve this work.

It is a great honor for me to take this opportunity to express my deep gratitude to Dr. Sherif El Dyasti, Assistant Professor, Electronics and Communication Department, College of Engineering and Technology, Arab Academy for Science, Technology and Maritime Transport (AAST), for his excellent cooperation, his expert help, continuous encouragement and valuable effort for completion of this work.

My special thanks and appreciation to Prof. Hazem Shatila, Virginia Polytechnic Institute and State University, Professor of Artificial Intelligence & Markovdata, CEO, thanks for spending his precious time and for his continuous encouragement that was behind the completion of this work.

Conflict of interest

"The authors declare no conflict of interest."

Appendices and nomenclature

AI	artificial intelligence
AMP	approximate message-passing
AoA	analysis of alternatives
AE	autoencoder
BER	bit error rate
CE	channel estimation
CNN	convolutional neural network
CS	compressive sensing
CR	compression ratio
CSI	channel state information
CsiNet	channel state information network
CsiNet-GRU	channel state information network-gated recurrent unit
DL	deep learning
DNN	deep neural network
FDD	frequency division duplex
GRU	gated recurrent unit
LMU	Legendre memory unit
LASSO	least absolute shrinkage and selection operator
LSTM	long short-term memory
MIMO	multiple-input multiple-output
mmWave	millimeter wave
MSE	mean squared error
NMSE	normalized mean square error
RELU	rectified linear unit
RNN	recurrent neural network
SNR	signal-to-noise-ratio

Author details

Hany Helmy[1*], Sherif El Diasty[2] and Hazem Shatila[3]

1 Artificial Intelligence, Python Developer and Data Scientist, Machine Learning Engineer, Cairo Airport Company (CAC), Cairo, Egypt

2 Department of Electronics, Arab Academy for Science, Technology and Maritime Transport (AASTMT), Cairo, Egypt

3 Department of Artificial Intelligence and Markovdata, Virginia Tech, Cairo, Egypt

*Address all correspondence to: hany.nabil@cairo-airport.com; hnabil110@gmail.com

References

[1] Liu G, Jiang D. 5G: Vision and requirements for mobile communication system towards year 2020. Chinese Journal of Engineering. 2016;**2016**:1-8. DOI: 10.1155/2016/5974586

[2] Rao X, Lau VKN. Distributed compressive CSIT estimation and feedback for FDD multi-user massive MIMO systems. IEEE Transactions on Signal Processing. 2014;**62**(12):3261-3271. DOI: 10.1109/TSP.2014.2324991

[3] Liu L et al. The COST 2100 MIMO channel model. IEEE Wireless Communications. 2012;**19**(6):92-99. DOI: 10.1109/MWC.2012.6393523

[4] Jiang Z, Chen S, Molisch AF, Vannithamby R, Zhou S, Niu Z. Exploiting wireless channel state information structures beyond linear correlations: A deep learning approach. IEEE Communications Magazine. 2019; **57**(3):28-34. DOI: 10.1109/MCOM.2019.1800581

[5] Xie H, Gao F, Jin S. An overview of low-rank channel estimation for massive MIMO systems. IEEE Access. 2016;**4**: 7313-7321. DOI: 10.1109/ACCESS.2016.2623772

[6] Cayamcela MEM, Lim W. Artificial intelligence in 5G technology: A survey, 2018. In: 2018 International Conference on Information and Communication Technology Convergence (ICTC), Jeju, Korea (South). pp. 860-865. DOI: 10.1109/ICTC.2018.8539642

[7] Helmy HMN, Daysti SE, Shatila H, Aboul-Dahab M. Performance enhancement of massive MIMO using deep learning-based channel estimation. IOP Conference Series: Materials Science and Engineering. 2021;**1051**(1):012029. DOI: 10.1088/1757-899X/1051/1/012029

[8] Diógenes GK, De Sousa Vitória AR, Silva DFC, Pagotto DDP, Sousa RT, Filho ARG. Live births prediction using legendre memory unit: A case study for the health regions of Goiás. In: 2023 IEEE 36th International Symposium on Computer-Based Medical Systems (CBMS); L'Aquila, Italy. 2023. pp. 329-334. DOI: 10.1109/CBMS58004.2023.00239

[9] Almamori A, Mohan S. Improved MMSE channel estimation in massive MIMO system with a method for the prediction of channel correlation matrix. In: 2018 IEEE 8th Annual Computing and Communication Workshop and Conference (CCWC), Las Vegas, NV, USA. 2018. pp. 670-672. DOI: 10.1109/CCWC.2018.8301699

[10] Perken ET. Spares channel estimation with regularization methods in massive MIMO systems. In: International Foundation Telemetring Conference Proceedings. 2018

[11] Donoho DL, Maleki A, Montanari A. Message passing algorithms for compressed sensing: I. motivation and construction. In: IEEE Information Theory Workshop on Information Theory (ITW 2010, Cairo), Cairo, Egypt. 2010. pp. 1-5. DOI: 10.1109/ITWKSPS.2010.5503193

[12] Wen C-K, Shih W-T, Jin S. Deep learning for massive MIMO CSI feedback. IEEE Wireless Communications Letters. 2018;**7**(5):748-751. DOI: 10.1109/LWC.2018.2818160

[13] Li X, Wu H. Spatio-temporal representation with deep neural recurrent network in MIMO CSI feedback. IEEE Wireless Communications Letters. 2020;**9**(5):653-657. DOI: 10.1109/LWC.2020.2964550

[14] Dong P, Zhang H, Li GY. Machine learning prediction based CSI acquisition for FDD massive MIMO downlink. In: 2018 IEEE Global Communications Conference (GLOBECOM), Abu Dhabi, United Arab Emirates. 2018. pp. 1-6. DOI: 10.1109/GLOCOM.2018.8647328

[15] Su X, Yu S, Zeng J, Kuang Y, Fang N, Li Z. Hierarchical codebook design for massive MIMO. In: 2013 8th International Conference on Communications and Networking in China (CHINACOM), Guilin, China. 2013. pp. 178-182. DOI: 10.1109/ChinaCom.2013.6694587

[16] Schwarz S, Heath RW, Rupp M. Single-user MIMO versus multi-user MIMO in distributed antenna systems with limited feedback. EURASIP Journal of Advanced Signalling Process. 2013; **2013**:54. DOI: 10.1186/1687-6180-2013-54

[17] Choi J, Chance Z, Love DJ, Madhow U. Noncoherent Trellis coded quantization: A practical limited feedback technique for massive MIMO systems. IEEE Transactions on Communications. 2013;**61**(12): 5016-5029. DOI: 10.1109/TCOMM.2013.111413.130379

Section 2

Current Applications and Future Trends

Chapter 2

Artificial Intelligence Techniques for Political Risk Management: An NLP Analysis of the 2019 US-China Trade War

Michael D. Wang

Abstract

This chapter examines the use of artificial intelligence (AI) techniques in natural language processing (NLP) for risk management, with a particular focus on applications in the field of political economics. The aim of this analysis is to identify and measure potential political risks by conducting a textual analysis of newspapers and social media, using sentiment scores as proxies for nationalism. The study uses the 2019 US-China Trade War as a natural experiment to evaluate the impact of international disputes on political risks. One significant finding is the positive effect of the trade war on sentiment in China's media about the US, which is attributed to the Chinese government's efforts to mitigate the negative impact of the trade war on international relations. The study also reveals a negative impact on bilateral imports due to the conflict. Furthermore, the study employs a Difference-in-Difference (DID) model to investigate the impact of news censorship on media during the trade war. It is found that China's regulators attempted to soften domestic anti-US sentiment, while the US media reported more negatively about China during the conflict. Overall, this analysis demonstrates how NLP technology can be effectively used to identify changes in the management of political risks by analysing news and other media.

Keywords: international conflicts, nationalism, international trade, international political economy, textual analysis

1. Introduction

McCombs and Shaw were the first to propose the agenda-setting function of mass media, which describes how media is used to affect the political positions of the recipients of information [1]. As newspapers are the main source of information in people's daily lives, McCombs and Shaw propose that the press plays a key role in influencing citizens' attention and even shapes their views on news topics [2]. For decades, some scholars have suggested that Chinese authorities use government-controlled media to maintain the status quo by shaping the public's opinions [3, 4].

McCombs and Shaw developed the agenda-setting theory by studying the 1968 presidential election in the US, which is a country with a relatively high degree of media freedom. The media sets agendas by appealing to people's emotions [1]. For example, the 2020 coronavirus disease outbreak has caused a global health crisis. Many people in China and other regions have died from the epidemic. On 3 February 2020, the *Wall Street Journal* posted an article entitled "China is the real sick man of Asia" and we can easily sense the negative and racist emotions towards China in this narrative [5]. News sentiment is the position of the media on the subject reported and may be set by authorities or driven by the tastes of readers. When a particular news sentiment is transmitted through the traditional local media, we can deduce that most of a relatively independent media are trying to satisfy a public figure, or that the authorities are setting the agenda for the mass media.

The connection between media and nationalism has been discussed in several works. Eriksen claimed that media such as the Internet promotes ethnocentric orientations among national and ethnic [6]. Hyun et al. indicate that user-generated content in media may facilitate actions of nationalism [7]. To figure out the causal effect of media on nationalism, sentiment could be a middle variable. One example is the anti-Japanese sentiment in China nowadays. Zhou and Wang suggest political propaganda has a significant effect on negative sentiment towards Japan [8]. For this example, political propaganda is the earliest textual data to appear on public platforms. These textual data with some elements led to a negative sentiment among readers and finally led to anti-Japanese political actions in China. Today's Chinese anti-Japan activities are a good example of Chinese nationalism, and the negative sentiment driven by political propaganda as a middle variable provides a channel for researchers to quantify the degree of nationalism.

Nationalism is a popular research topic for political scientists. A large number of studies focus on the causes of nationalism and its political influence on conflict or democracy, among other issues [9–13]. However, there is relatively little quantitative research comparing nationalism in different countries or discussing the role of the authorities in promoting or supressing nationalism in international conflicts. Nationalism can have positive and negative effects on bilateral conflicts. During the South China Sea dispute, the Philippines' attempt to end Chinese fishermen's illegal poaching was stopped by a Chinese maritime reconnaissance vessel, intensifying the conflict on 8 April 2012. The *Global Times* (affiliated with the *People's Daily*) followed up on a statement by the Vice Foreign Minister of China and reported that China would not rule out the possibility of using force to resolve the conflict with the Philippines [14]. In contrast, during the Diaoyu (Senkaku) Islands dispute in 2012 between China and Japan, Fravel argues that there was no obvious trend in the frequency or timing of articles about the issue published by the *People's Liberation Army Daily* or the *People's Daily* (two official newspapers representing the attitudes of China's military and government), suggesting that China avoided mobilising public opinion when resolving the dispute [15]. In these two conflicts, the Chinese government's approach to media censorship and its attitude towards nationalism were different. Storey proposes that the weakness of the Philippine navy gave the People's Republic of China an opportunity to extend its sovereignty in the South China Sea [16]. We can infer that military power is part of the bargaining power of a party in an international conflict, so whether or not this power is invoked can affect the positive and negative effects of nationalism on the achievement of desirable negotiation outcomes.

In this chapter, I measure the level of nationalism in English and Chinese regions using sentiment consistency in the mainstream press. I also discuss the role of the news censorship system in China during the 2019 US-China trade war and compare it with the press in the US, which has a higher degree of media freedom. To study the strategies that governments adopt to manage nationalism during international conflicts, it is first necessary to find an indicator that accurately measures nationalism. I create a news consistency indicator based on the textual analysis of newspapers to capture the degree of emotional deviation between newspapers. After controlling for media bias and other macro factors, this news consistency indicator can be used to measure the level of nationalism. In addition, to examine government interference in news content, I use the 2019 US-China trade war as a natural experiment to identify the role of news censorship on nationalism and the effects of nationalism on the relations between the two parties engaged in a dispute. Identifying the position of the negotiating parties on nationalism can provide a means of determining the relative bargaining power of the parties and the possible outcome of a negotiation.

This chapter is organised as follows. Section 2 reviews the literature on nationalism and news censorship. We also discuss previous studies of the outcomes of trade wars to clearly explain the context of my natural experiment. Section 3 describes how I construct my data set to measure nationalism and relations between the US and China. Section 4 presents my approach to transform my data into metrics. Section 5 indicates the conceptual framework and key variables for empirical analysis. Section 6 presents my main results on how news censorship affects US-China relations at the time of the trade war. Section 7 tests the robustness of my main results and in Section 8 I offer suggestions.

2. Literature review

2.1 Nationalism

Nationalism is an important factor in the economy, although economists often overlook it. Breton shows that political nationalism can influence countries' capital investment in resources in nationality or ethnicity by household, business and government [9]. It is important to first define nationalism. Hayes [17] divides nationalism into four types:

> *"It stands in the first place for an actual historical process, that of establishing nationalities as political units, of building out of tribes and empires the modern institution of the national state. Secondly, the term indicates the theory, principle, or ideal implicit in the actual historical process; in this sense it signifies both an intensification of the consciousness of nationality and a political philosophy of the national state. Thirdly, it may mean, in such phrases as 'Irish nationalism' or 'Chinese nationalism', the activities of a particular political party, combining an historical process and a political theory; this meaning is clearer when the adjective 'nationalist' is employed, for example, in speaking of the historical Irish Nationalist Party. A fourth and final use of 'nationalism' is to denote a condition of mind among members of a nationality, perhaps already possessed of a national state, a condition of mind in which loyalty to the ideal or to the fact of one's national state is superior to all other loyalties and of which pride in one's nationality and belief in its intrinsic excellence and in its 'mission' are integral parts."*

Hayes uses the fourth definition to discuss the relationships between nationalism and economic theory, arguing that nationalism will perpetuate and exacerbate the difference between the natural sciences and the social sciences [12]. I follow the same (the fourth) definition of nationalism in the main sections to quantify and analyse the state of mind of the members of a nation. China provides an ideal case for researchers to capture nationalism among citizens based on the media activities of the Chinese Communist Party (CCP), because the CCP plays a key role in the political life of Chinese people.

Although each country has a different level of nationalism, many scholars and politicians pay special attention to Chinese nationalism because of China's important position in the economy and military affairs of East Asia and its geopolitical influence on neighbouring countries. Coenders and Scheepers reveal the mechanism that links nationalism and national economic conditions by focusing on Europe during the 19th century [10]. They describe nationalism as a tool the elite use to persuade the public to accept the established social order after industrialisation. First, nationalism helps maintain the current social order. Second, it can contribute to the short-term growth of the domestic economy. Heilmann estimate the effects of international conflicts on bilateral trade relations using several politically motivated boycotts and conclude that boycotts of consumer products are the most effective, especially of well-known brands [18]. From the authorities' point of view, a fall in demand for exports can partly translate into domestic demand and stimulate economic growth and create jobs. However, in particular situations, such as when there is political corruption or unequal international conflicts, nationalism can become an obstacle to authoritarian social stability and the progress of negotiations. Kuzio uses Ukraine's "Orange Revolution" in 2004 as a case study to illustrate the positive effects of nationalism on the democratisation of post-communist countries [11]. From the authorities' point of view, nationalism was not beneficial in maintaining the status quo of communist Ukrainian society.

In this chapter, I study the positive and negative effects of nationalism by analysing the bilateral relations between the US and China during the 2019 US-China trade war.

2.2 Trade disputes

When Donald D. Trump was elected President of the United States he officially raised tariffs on Chinese exports, starting the 2019 US-China trade war. Liu and Woo summarise three major concerns leading to the decision to launch a trade war against China: (i) China's chronically large trade surplus slows job creation in the US; (ii) China's illegal and unfair access to US intellectual property; and (iii) China's effect on the US-led international system and its own national security [19]. Although the US launched a trade war to defend its economic interests, different scholars have different opinions on the effect of this trade war. Li, He and Lin argue that the US has more to gain than China in the trade war negotiations based on numerical stimulation results and therefore claim that the US has more bargaining power than China [20]. Conversely, using evidence of the consumption effects of trade shocks, Waugh suggests that China's retaliatory tariffs will lead to concentrated welfare loss in the US [21]. A more general argument is that neither China nor the US will win in the trade war, only cooperation can lead to a win-win situation. For example, Berthou et al. conclude that nobody wins a trade war and that global real GDP generally decreases by up to 3% after two years based on a multi-region dynamic general equilibrium model [22].

The trade war provides a natural experiment to identify the effects of nationalism on an economic conflict between the two largest economies in the world. However, an important question related to the study of nationalism is how to measure it accurately. As a state-to-state relationship is a relatively abstract concept, in this study I use textual data from newspapers to quantify nationalism. Indeed, nationalism and the mass media are tightly bound together.

2.3 Media and censorship system

Montiel et al. investigate the production of nationalism in the national media during an international conflict [23]. Nationalism also has an effect on the public media. Studying the India-Pakistan military conflict of 29 September 2016, Pandit and Chattopadhyay argue that the news media in India have always adhered to belligerent reports of patriotic nationalism and consciously obscured the disagreement of minority voices [24]. The strong correlation between media reporting and nationalism allows us to use the media as a proxy for measuring nationalism. Although using news articles allows us to measure nationalism, another problem in this measure is the effects of a news censorship system used by some governments.

China's news censorship system limits media coverage of information deemed harmful to the stability of its regime. In general, this type of media censorship is seen as part of authoritarian control of information to maintain the power of the regime and increase control over public sentiment. It is generally believed that authoritarian regimes inevitably limit media independence. For instance, Lorentzen argues that China strategically adjusts the number of reports allowed based on potential social tensions [25]. King et al. use textual analysis of social media to compare content before and after the government imposes censorship to show that censorship is designed to prevent current collective activity or activity that may occur in the future [26]. From the Chinese authorities' point of view, censorship is a means of ensuring that news content is in the interest of the government. Therefore, by comparing the differences in the effects of nationalism on the US and China during the natural experiment of the trade war, it is possible to identify the effect of the media censorship system. In addition, we can determine whether nationalism sparked by the trade war is in the interest of the government.

3. *Factiva* news and trade data

In this section, I present the main data sources I use to create meaningful indicators of US-China relations, and focus on two measures: (i) the sentiment score of news articles and (ii) bilateral imports between the US and China. For the first measure, I use the *Factiva* database to find and collect raw newspaper articles that interest us. I also apply a sentiment analysis to process the textual data and extract useful quantitative information. For the second measure, I use the annual bilateral import data from the WTO. I discuss in detail our data sources and processing methods in the next section.

Factiva is a database provided by Dow Jones & Company that allows researchers to search for global textual data, such as newspapers, journals and magazines. I use this database to collect textual data from newspapers to generate my text-based indicators.

Based on their circulation, I select the *Wall Street Journal*, the *New York Post*, *The Boston Globe*, the *Star Tribune*, *USA Today*, *The New York Times* and *The Washington Post* to construct my US news corpus, and use “China” as the keyword to find articles related to economic, financial and political topics before 5 December 2019. For the Chinese news corpus, I choose various newspapers including *Ta Kung Pao*, *Wen Wei Po*, the *Global Times*, *People's Daily Online*, the *Sing Tao Daily* and *The Beijing News*, and use “<001> “(the US) as the keyword.

One challenge in cleaning the raw *Factiva* data is that the original data format downloaded from the database is in either rich text format (rtf) or portable document format (pdf). To perform textual analysis on these unstructured data, I need to convert them to comma-separated values (csv) files by date, source, title and content columns. For this data cleaning task, in the first step, I use Python codes to automatically convert textual data from pdf to txt format using “utf-8” (as some of my textual data are in Chinese, the encoder “utf-8” instead of “ascii” is required to process non-English characters). Next, I use regular expressions and the name of each newspaper to detect one complete news article from the txt data files and match the date, newspaper, title and content of each news article.

Sentiment is a qualitative value connected to people's emotions, subjective cognition and psychological activities. To quantify sentiment in news articles, I use a dictionary approach that counts the frequency of positive and negative terms of emotional arousal (descriptions such as “amazing”, “happy”, etc.) based on the *National Taiwan University Semantic Dictionary* (NTUSD).[1] The NTUSD uses a machine learning approach to detect and collect positive and negative terms from English and Chinese documents, which can be applied to sentiment analysis on textual data of general topics.

The WTO Data portal provides statistical indicators, including bilateral imports by detailed product or service sectors from 1948 to the present. In this study, I use import data from 2002 to 2019 after China joined the WTO. Imported products and services are classified into 97 sectors, from live animals to works of art. I accumulate the US dollar value of all sectors in one year and use annual imports as my second measure of US-China relations.

4. Sampling methodology

4.1 Preprocessing for textual analysis

In this section, I present the methodology I apply to convert the large amount of news articles generated in Section 3 into more useful data by removing noise in the texts. To empirically analyse the 64,026 English news articles and 123,549 Chinese news articles in my two corpora, one major challenge is to develop appropriate measurement approaches. Before analysing the news articles, I preprocess the raw content in several steps according to the language used. For the US news corpus, the first step of preprocessing is to remove stop words such as “a” and “the” that commonly appear in English articles. Numbers and punctuation are also removed from the raw content during this step. In the second step, I convert the remaining English words to their linguistic roots, for instance, the token “save” can indicate “saving”, “savings” and

[1] Available at: http://nlg.csie.ntu.edu.tw

"saved". In the last step, I filter the remaining terms using TF-IDF proposed by Hansen, McMahon and Prat [27]. **Figure 1a** shows the TF-IDF weight of each English term. I eliminate all of the tokens ranked 100,000 or lower.

For the Chinese news corpus, the first step of preprocessing is to divide all of the Chinese texts into meaningful words. In the second step, I delete all of the Chinese stop words, numbers and punctuation, as with the US news corpus. As there is no linguistic root for each Chinese word, I skip the stemming process in the preprocessing of the Chinese news corpus. Instead, I go directly to the last step and filter the remaining terms based on the tf-idf weights of each word. **Figure 1b** shows the weight of each Chinese term. I remove all of the words ranked 200,000 or lower after inspection.

4.2 News sentiment

To interpret the opinion of each article in the US (Chinese) news corpus with regard to China (the US), I retrieve emotional information from news articles using the sentiment score. One challenge for analysing a large-scale news corpus is to design an appropriate and concordant algorithm to capture the sentiment score for each document written in different languages. Moreover, there are many more news sources available in the world than my selected newspapers providing new information at different times, so efficiency is another important criterion for the sentiment analysis algorithm. Based on these considerations, I apply the dictionary approach in textual analysis by calculating the total frequency of each word of emotional arousal based on a pre-established positive (negative) dictionary. To compute the absolute sentiment score (Σ) for article i at time t, I use the number of positive words (p) minus the number of negative words (n) divided by the total number of terms of emotional arousal ($p + n$) as follows:

$$\Sigma_{it} := \frac{p_{it} - n_{it}}{p_{it} + n_{it}}. \quad (1)$$

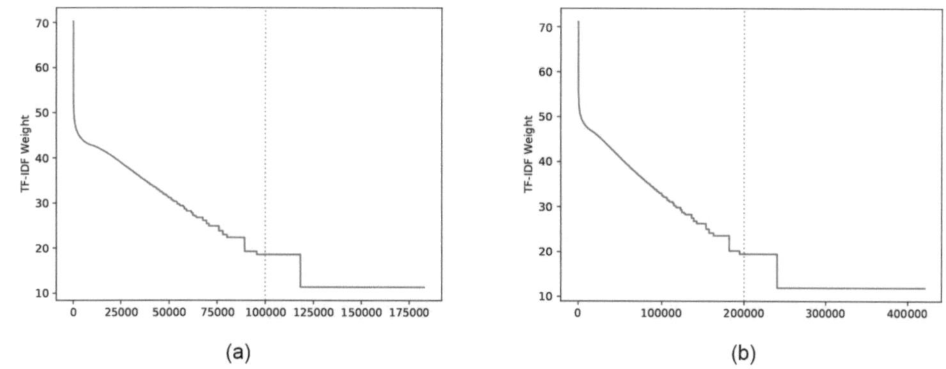

Figure 1.
Rank of Terms Ordered by TF-IDF. Note: *For each term t, the term frequency is $tf_t := 1 + \log(n_t)$ where n_t is the count of term t in the corpus. The inverse document frequency is $idf_t := 1 + \log\left(\frac{1+N}{1+df_t}\right)$ where N is the total number of documents in the corpus and df_t is number of documents where the term t appears. The product of tf_t and idf_t is the weight of tf-idf weight of term t denotes the informativeness of t in the corpus. (a) US news corpus, (b) chinese news corpus.*

Σ is scaled continuously within the range from −1 to 1. For instance, if the number of positive words in an article is 10 and there is no negative word, Σ is $(10-0)/(10+0)=1$. The annual average absolute sentiment of each newspaper from the US and Chinese sources for the years 2009 to 2019 is presented in **Figures 2** and **3** visualises the sentiment in a heatmap.

When an article does not contain any positive or negative words, I represent this scenario with the symbol Σ as "nan" (not a number). One typical strategy for handling

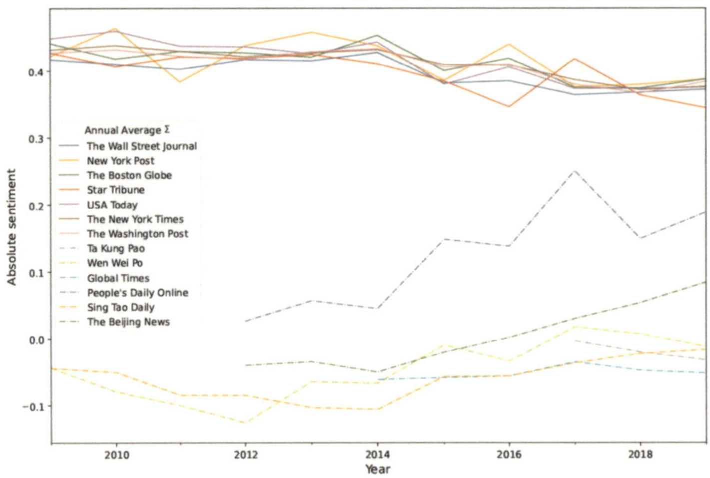

Figure 2.
Annual absolute sentiment of newspapers.

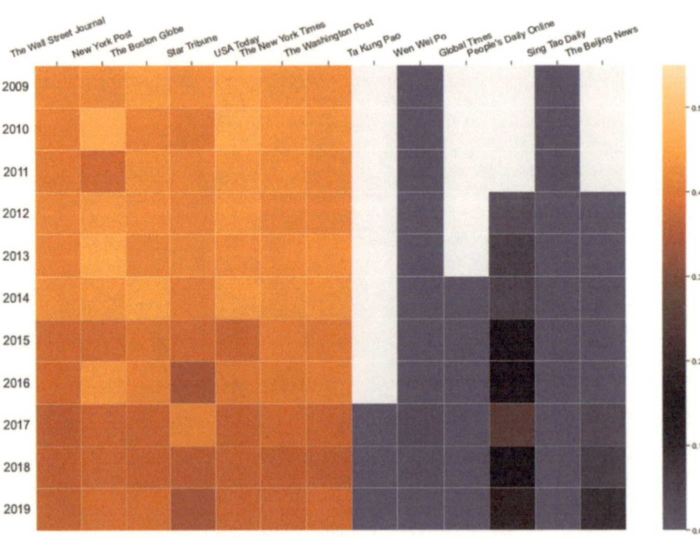

Figure 3.
Annual sentiment of newspapers. Note: This heatmap presents the annual average sentiment of each newspaper from the US and Chinese sources. For each year within the row, the colour and depth of each shading indicates the polarity and degree of the sentiment score for each newspaper within the column. Specifically, orange (blue) means the domestic media on average more positively (or negatively) reports the articles of the counter-party. Relative sentiment score is the ratio between the absolute sentiment score and the market average score and then minus one, which can be applied to measure robust relationships between different textual data sources regardless the language they used.

such "nan" values is to treat them as neutral sentiment. However, it should be noted that in my corpus, such "nan" values are a rare occurrence, accounting for only approximately 0.81% of the data. Consequently, I have chosen to exclude these observations from my analysis as part of an effort to streamline the model. Nonetheless, I believe that this exclusion will not significantly compromise the validity of my conclusions.

4.3 Topic model

The original database provides a general category for each news article and I focus on all articles marked as "economic", "financial" and "political". However, I am interested in specific topics and in the popularity of these topics among news media recipients. To identify the latent topics of each document, I use the Latent Dirichlet Allocation (LDA) topic model to process the textual data. As Baker, Bloom and Davis discussed, one challenge in LDA topic modelling is to choose the appropriate number of topics K [28]. Typically, a higher K increases the statistical capability of the prediction model, but sacrifices its interpretability, as each topic may be too specific to be analysed. In the extreme case where $K = D$, each document will be assigned one individual topic. Although this model fits the sample perfectly, it is useless in my case to capture the general patterns in each text, and vice versa. To resolve the trade-off between a high K and a low K, I compare the experimental results of model perplexity[2] in a 20% validation data set with different K values, as shown in **Figure A.1**. In the end, I use $K = 850$ for the textual data in English and $K = 550$ for the Chinese documents by taking into account the trade-off between the complexity of the model and the validation perplexity. I also report my empirical results for $K = 900$ for the US news corpus.

5. Analysis methodology

In this section, I describe this study's conceptual framework and explain which set of variables I chose for analysis. **Figure 4** denotes the theoretical framework with a causal graph, and the critical variable is the media. I define media as the textual data in newspapers, the Internet, or other public platforms. These textual data can convey much information, and I mainly focus on the popularity of news topics and media sentiment on them. Censorship can be implemented under the control of the regulator, and the gap between authoritative and democratic governments also affects media consistency differently. I define media consistency as the concentration of sentiments among different media platforms. Consistency depends on the current media sentiment, as well as the interaction between censorship and topic popularity. One difficulty is to observe the censorship variable since regulators may not willing to make these data observable to the public. To address this problem, I assume the effect of regulator is relatively stable, however national conflicts such as the trade war (t-war) can shift the degree of censorship. The theoretical framework between media or sentiment and nationalism is clear from previous studies. The added literature indicates the relationships between media and sentiment [8] and the relationship between media and nationalism [6, 7].

---

[2] In information theory, perplexity is an indicator used to measure how well a probability model predicts a sample. Usually, a low perplexity indicates that the probability distribution is better able to fit the sample.

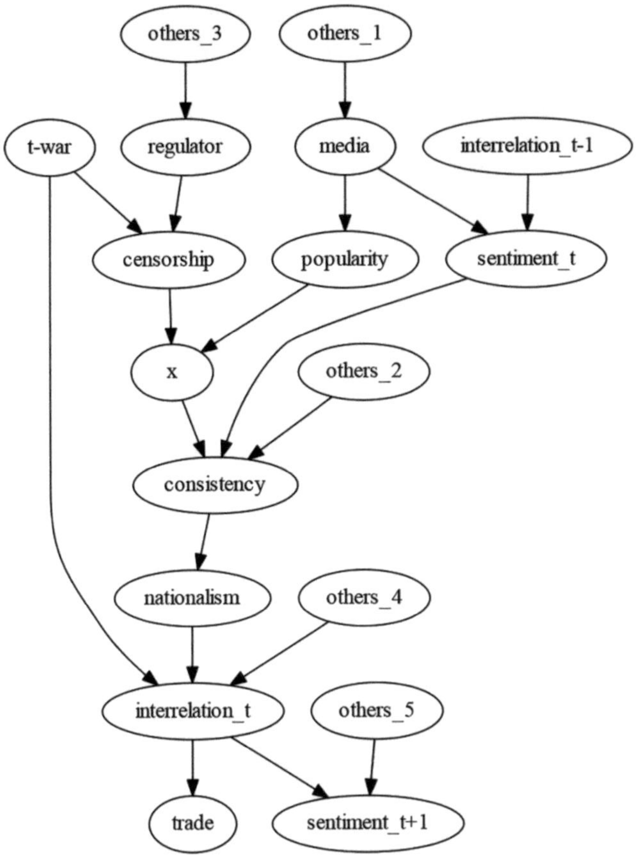

Figure 4.
Theoretical Framework with Causal Graph. Note: *I draw a causal system as a directed graph to denote all causes of the outcome of interest in this study, based on past literature and assumptions grounded in theory and empirical findings. Each graph node indicates a variable, and single-headed arrows represent causal effects. The parent of the directed edge leads to the variable at the child's position. Variable x denotes the interaction effects of popularity and censorship on media consistency.*

5.1 From trade war to interrelation

The basic model I build to study the effects of the trade war on the relations between the US and China is as follows:

$$y_{it} = \beta_0 + \beta_1 D(T - war)_t + \gamma \mathbf{X}_t + \alpha_i + \varepsilon_{it}, \tag{2}$$

where y_{it} is a dependent variable indicating the measures of US-China relations for entity i at time t. I use lagged media sentiment as a proxy variable of the degree of national relations. The causal path can be interpreted by the previous national relations affecting the government's attitude towards regulating the media, as well as the sentiment in the media narrative. As a result, current media sentiment can be used to measure national relations approximately. The model incorporates a fixed effect for each entity, denoted by α_i, and a constant term, denoted by β_0. The parameter of interest, β_1, is estimated to examine the impact of the trade war on national relations. Additionally, the vector of macro-level control parameters is denoted by γ. I employ panel data regression with entity fixed effects to estimate the model parameters.

Table 1 denotes the descriptive statistics of the effects of the trade war on national relations. Sentiment_{t+1} is the proxy of national relations between China and the US, and I set it as the dependent variable. *D(T-war)* is a dummy variable indicating whether the date of publication is during the US-China Trade War (1 after May 2016, and 0 before), and vector $\boldsymbol{X}_t$ controls the political, financial and other effects related to y_{it}. For the political control, I include the Chinese Political Uncertainty (CPU) index discussed by [29]. The CPU index is a text-based index focusing on specific terms related to uncertainty in US-China bilateral relations expressed in the *People's Daily*. As mentioned above, the *People's Daily* started publishing in 1948 as an official publication of the CCP and reflects the position of the Chinese government. Both the CPU index and my news sentiment measures use a dictionary approach that considers the frequency of terms or the number of documents containing specific words found in a pre-established dictionary. In particular, the CPU index captures information on public uncertainty about national policy and the economy, which can be considered as a quantitative proxy for people's perception of the monthly level of systematic risk in national relations. For the financial control (Market Return), I consider the stock market returns on the day before the publication of a given article. It is easy to imagine that people's short-term gain or loss in a financial market can influence their emotions. For the US market, I use the S&P 500 market index as a proxy for the performance of the US financial market. For China, I use the SSE Composite Index to reflect the performance of the Chinese financial market.

In addition, I include certain specific controls to measure US-China relations. For the news sentiment measures, whether a news topic is popular or not can influence the writers' use of words that trigger emotional arousal. If a publisher agrees with the tastes of the public, this publisher will be more motivated to make a positive or

Variables	Obs.	Mean	Median	Std. Dev.	Min	Max	Remark
US							
Sentiment_{t+1}	42,144	0.401	0.404	0.119	−1	1	dep
D(T-war)	42,145	0.401	0	0.49	0	1	indep
CPU Index	42,145	148.405	139.035	47.237	63.877	284.136	indep
$\text{MarketReturn}_{t-1}$	42,145	0.000265	0.000697	0.0105	−0.0666	0.0707	indep
D(Popular)	42,145	0.577	1	0.494	0	1	indep
D(Unpopular)	42,145	0.110	0	0.313	0	1	indep
China							
Sentiment_{t+1}	81,861	−0.0119	−0.0303	0.444	−1	1	dep
D(T-war)	81,862	0.552	1	0.497	0	1	indep
CPU Index	81,862	525.577	403.254	423.448	160.821	2787.628	indep
$\text{MarketReturn}_{t-1}$	81,862	0.000277	0.000645	0.0137	−0.0849	0.0612	indep
D(Popular)	81,862	0.467	0	0.499	0	1	indep
D(Unpopular)	81,862	0.208	0	0.406	0	1	indep

Notes. This table reports the descriptive statistics of variables in my observations. I use "dep" to denote the dependent variable and "indep" which means the independent variables in my model.

Table 1.
Descriptive statistics of the trade war on national relations.

negative description. Conversely, if a publisher is seeking to manipulate public sentiment by using emotionally arousing language, popular and unpopular news topics are not the ideal target. On the one hand, overly popular topics can easily lead to widespread discussion and expose the purpose of the publisher. On the other hand, unpopular topics do not arouse the interest of readers, leading to failed manipulation. Accordingly, I add a topic-specific control for the news sentiment measures using two dummy variables, $D(Popular)$ and $D(Unpopular)$, indicating whether a news-based item covers a popular or unpopular topic, respectively. I classify a news article discussing a latent topic with a popularity index (see Section 6.3 for more details) in the top 35% as popular news ($D(Popular) = 1$, and 0 otherwise). Conversely, I define an article as unpopular if its topic has a popularity index in the bottom 35% ($D(Unpopular) = 1$, and 0 otherwise). This control allows us to identify the effect of readers' interest on how the news is presented.

Table 2 denotes the descriptive statistics of the effects of the trade war on bilateral imports between the US and China from 2017. On the one hand, the direct effect of the trade war on bilateral trading is an interesting study. On the other hand, bilateral trading could approximately be a proxy variable of national relations besides the lagged media sentiment. Therefore, the two different proxy variables of national relations provide a robust estimation of the impact of the trade war on bilateral relations.

For the measure of bilateral imports, I add two controls, the total annual population (Population) and the monthly number of WTO dispute cases between the US and China (D(Dispute)), instead of the topic controls. The population control (unit: billion) allows us to capture the effect of the natural increase in people's demand for foreign products and services. The WTO dispute case dummy $D(Dispute)$ controls the effects of other bilateral trade disputes besides the US-China trade war on people's demand for the other party's products and services. $D(Dispute) = 1$ if there is one or more WTO dispute cases between the US and China, and 0 otherwise.

Finally, I add a source fixed effect α_i to represent the heterogeneity of each newspaper (for the sentiment measures) or of each country (for the import measure). For the

Variables	Obs.	Mean	Median	Std. Dev.	Min	Max	Remark
US							
Import	36,428	0.355	0.420	0.161	0.0	0.498	dep
D(T-war)	36,428	0.265	0	0.442	0	1	indep
$MarketReturn_{t-1}$	36,428	0.000205	0.000575	0.0107	−0.0666	0.0708	indep
Population	36,428	0.318	0.318	0.00680	0.306	0.327	indep
D(Dispute)	36,428	0.195	0	0.396	0	1	indep
China							
Import	70,113	0.105	0.134	0.0622	0.0	0.153	dep
D(T-war)	70,113	0.406	0	0.491	0	1	indep
$MarketReturn_{t-1}$	70,113	0.000221	0.000652	0.0137	−0.0849	0.0612	indep
Population	70,113	1.373	1.379	0.0174	1.331	1.393	indep
D(Dispute)	70,113	0.171	0	0.377	0	1	indep

Table 2.
Descriptive statistics of the trade war on imports.

sentiment measures, one major driver of heterogeneity is media bias. Different newspapers may have a specific and consistent tendency to subjectively report news in a way that favours certain institutions, organisations or political parties. Other possible reasons for heterogeneity include the linguistic characteristics of the articles and the demographic characteristics of the target customers. Articles in different languages and from different regions may use sentimental terms at different frequencies. For the import measure, each country's national import demand will lead to heterogeneity. My main concern is the estimated coefficient β, which indicates the average change in sentiment across all entities in a region before and after the trade war event.

5.2 From censorship to nationalism

I use media consistency as a proxy of nationalism. Sentiment consistency in newspapers is an important factor that I need to take into account when explaining the dynamics of daily emotions in different newspapers. Newspapers may naturally or forcibly focus on the common voice of some or entire public groups due to sponsorship or censorship by a third party. On the one hand, news consistency depends on the current media sentiment. On the other hand, I measure news consistency by aggregating sentiment among massive media platforms and under the pressure of the media censorship system. In other words, news consistency should better interpret generalised nationalism than traditional media sentiment. Therefore, I choose news consistency as an instrumental variable of nationalism. For each country, I define a proxy for sentiment consistency $c_{it} \in]0, 1]$ in newspaper i at time t as follows:

$$c_{it}:=\frac{1}{1+|\Sigma_{it}-\overline{\Sigma}_t}|, \qquad (3)$$

where Σ_{it} is the average sentiment of newspaper i at time t, as discussed in Section 4.2, and $\overline{\Sigma}_t$ is the mean value of all news articles at time t. If c_{it} is close to 1, it means that the sentiment of newspaper i is more likely to be consistent with the common voice of the public media, and vice versa. In **Figures 4** and **5**, I use two histograms of variable c_{it} showing the frequency distribution of selected observations from all newspapers in my US news corpus (5a) and Chinese news corpus (5b) before and after the 2016, which is the year of the Trade War.

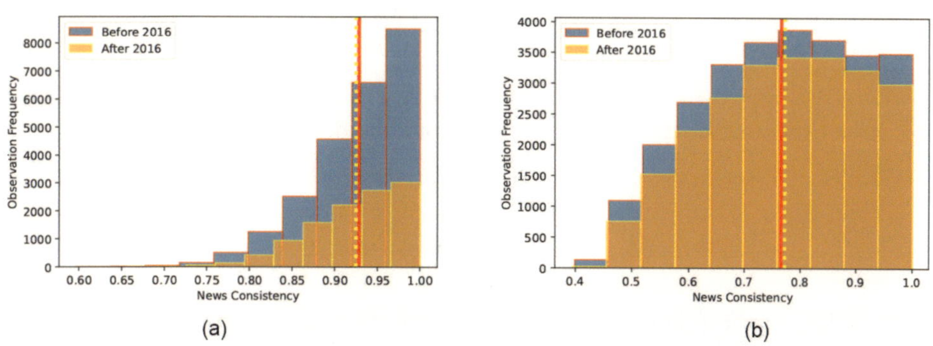

Figure 5.
News Consistency Histogram (c_{it}). Note: *These histograms show the frequencies of selected observations in my corpus of variable c_{it} in the US and China. The yellow dot indicates the average media consistency after 2016, whereas the red line denotes the average consistency from 1989 to 2016. (a) The United States, (b) China.*

Next, I discuss the characteristics of news consistency using a linear model as follows:

$$c_{it} = \beta_0 + \beta_1 \rho_{it} + \beta_2 D(T - war)_t \times \rho_{it} + \boldsymbol{\beta_x} \mathbf{X}_t + \varepsilon, \tag{4}$$

where ρ_{it} is the corresponding average popularity index of newspaper i at time t. To compute ρ_{it}, it is crucial to have knowledge of the popularity index of each topic to which the newspaper pertains. It is evident that some news topics are more appealing than others, and therefore people may show more interest in discussing them. As the topics produced by the topic model are not arranged in any specific order (see Section 4.3), I establish a popularity index for each topic by ranking them. The initial step in developing this index involves constructing distinct social network corpora for users who speak English and Chinese.

I collected a total of 15,236,749 tweets from the English corpus between 10 October 2018 to 8 August 2019 using the *Twitter* API. The specific keywords used to query the API were "tradewar", "tradewars", "MAGA", "trade", and "USChina". Subsequently, a subset of 30,473 unique narratives was constructed by random selection. It is assumed that the chosen tweets related to the 2019 US-China trade war encompass current economic and political narratives, which are of interest for this study. For the Chinese corpus, I utilised the *Weibo* public timeline API, a popular social networking platform in China similar to *Twitter*, to collect 3,194,226 unique narratives from the general population between May 29, 2018, and October 17, 2018, although not strictly in real-time. Subsequently, I chose 9539 unique comments to create my subset. Unlike the *Twitter* API, the *Weibo* API does not offer a keyword query feature to the public users. Therefore, I collected data on general topics to build my Chinese corpus.

In the second step of the analysis, I construct a p by q matrix $\boldsymbol{T}$, where p denotes the number of topics selected in the topic model and q represents the top q terms generated by the model that are of interest. For the primary analysis, I set $p = 850$ for the English corpus, $p = 550$ for the Chinese corpus, and $q = 12$. $\boldsymbol{T}_{\mu\nu}$ is utilised to capture the count of words in row μ and column ν in **Figure 6** or **Figure 7**, as presented in Section 6.1, for each social network corpus. Next, I compute the popularity score for each topic μ,

$$\boldsymbol{P}_\mu := \sum_{\nu=1}^{q} \boldsymbol{T}_{\mu\nu} \times \boldsymbol{W}_{\mu\nu}, \tag{5}$$

where $\boldsymbol{W}_{\mu\nu}$ represents the estimated probability that a word in column ν belongs to topic μ, as generated by the topic model. It should be noted that $\boldsymbol{P}$ is a column vector with p dimensions. To obtain the popularity index vector, $\boldsymbol{P}$ is normalised.

$$\boldsymbol{P}^* := \frac{\boldsymbol{P}}{\|\boldsymbol{P}\|_2} = \left[\rho_1, \rho_2, \ldots, \rho_p\right]^{\mathrm{T}}, \tag{6}$$

where $\|\boldsymbol{P}\|_2$ is the l^2-norm of $\boldsymbol{P}$, and the average popularity index ρ_i at time t can be expressed as

$$\rho_{it} = \frac{\sum_{x \in \boldsymbol{T}_x} \rho_x}{N_{it}} \quad \text{s.t.} \quad \mathbf{card}(\boldsymbol{T}_x) = N_{it}, \rho_x \in \boldsymbol{P}^*, \tag{7}$$

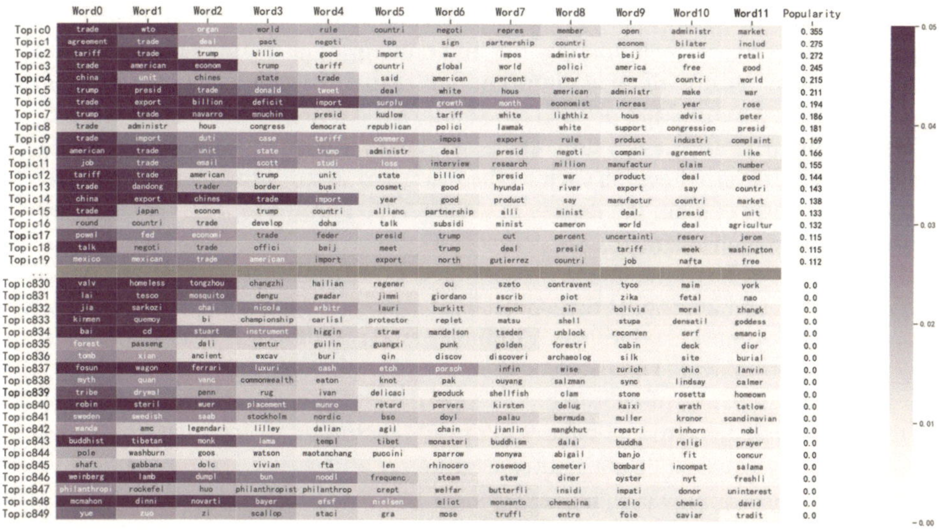

Figure 6.
Topics in the US News Corpus Ranked by Popularity. Note: *This figure denotes the 850 separate distributions over news vocabulary that LDA model learns to indicate topics. According to a popularity index that captures term frequencies of the top 12 vocabulary terms in each LDA topic in social networks, I order these distributions from 0 (the most popular) to 849 (the least popular).*

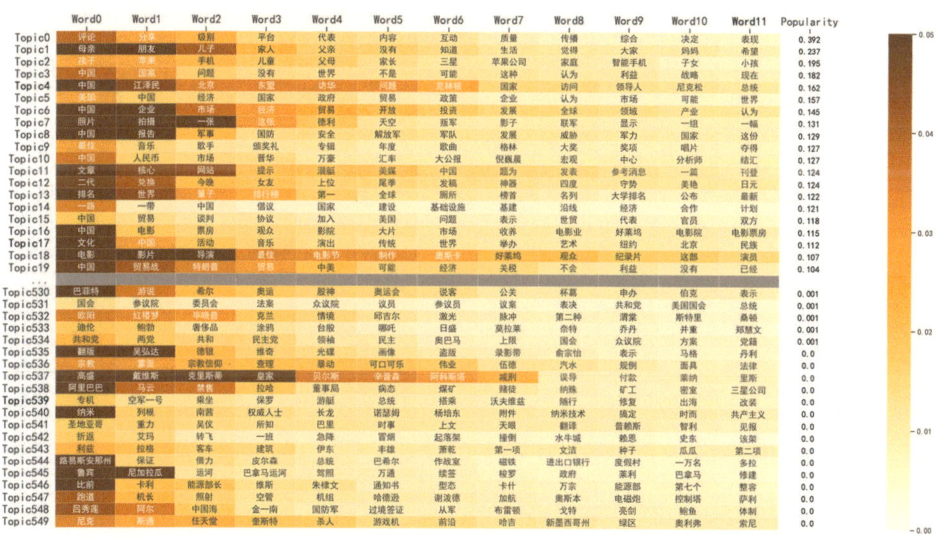

Figure 7.
Topics in the Chinese News Corpus Ranked by Popularity. Note: This figure denotes the 550 separate distributions over news vocabulary that LDA model learns to indicate topics. According to a popularity index that captures term frequencies of the top 12 vocabulary terms in each LDA topic in social networks, I order these distributions from 0 (the most popular) to 549 (the least popular).

where N_{it} represents the total number of articles in newspaper i at time t, $\boldsymbol{Tx}$ maps the set of articles $d_1, d_2, \ldots d_{N_{it}}$ to their corresponding topic categories generated by the LDA algorithm, and $\mathbf{card}(\cdot)$ denotes the cardinality of a set. The elements ρ_x belong to the vector $\boldsymbol{P}^*$, which contains the popularity index of each topic x.

$D(T - war)_t$ is the dummy variable for the US-China Trade War period, as discussed in Section 6.2. I assume the parent variables of consistency include the current media sentiment, the interaction between topic popularity and media censorship, and other factors, including politics and demographic characteristics. I add this interaction term because of an assumption that regulators may have stronger censorship over the news with high popularity, so there is a potential interaction effect between popularity and censorship. Once regulators' impact on censorship is relatively constant, external events such as the trade war as an instrumental variable should sufficiently explain the degree of censorship. **Table 3** reports the descriptive statistics of news consistency.

As in **Figure 4**, I assume media censorship only depends on two factors: one is regulators' will, and another is the external interrelation shock, such as the Trade War's impact on Sino-US relations. Within my study period, it is reasonable to consider the regulators' will is relatively stable, therefore I recognise it as a constant term. As a result, the dynamic of censorship meanly depends on the Trade War effect. Vector $\boldsymbol{X}_t$ controls for regional factors in the US, including political bias in the media[3] (using a dummy equal to 1 for a right biased source and 0 for a left biased source) and the proportion of Chinese people (excluding Taiwanese) in the state.[4] Notice that the Chinese ratio is only suitable in the US region. I do not create the US ratio because China is not an immigration country, and the percentage of US immigrants in China is trivial. In addition, I do not use the Chinese ratio in my sentiment regression model because I assume the sentiment variable depends on the media and previous international relations per se. This is a reasonable assumption because the media's primary purpose is to report new information regardless of demographic interests, especially since my sample sources are well-known US newspapers. However, I consider the Chinese ratio could affect the media consistency because a higher number of Chinese clients might lead to lower media consistency to report Chinese if the US media generally tend to report Chinese news negatively but still care about the local Chinese immigrants.

Variables	Obs.	Mean	Median	Std. Dev.	Min	Max	Remark
US							
Consistency	42,145	0.924	0.935	0.0553	0.465	1	dep
D(T-war)	42,145	0.401	0	0.490	0	1	indep
Popularity	42,145	0.0309	0.00776	0.0594	0.0	0.355	indep
D(Political Bias)	42,145	0.459	0	0.49800	0	1	indep
Chinese Ratio	42,145	0.0280	0.0328	0.00918	0.00512	0.0328	indep
China							
Consistency	85,617	0.771	0.778	0.135	0.388	1	dep
D(T-war)	85,617	0.572	1	0.495	0	1	indep
Popularity	85,617	0.0321	0.0195	0.0356	0	0.392	indep

Table 3.
Descriptive statistics of news consistency.

[3] Data source: *Media Bias/Fact Check*, https://mediabiasfactcheck.com.

[4] Data source: U.S. Census Bureau, 2011–2015 American Community Survey 5-Year Estimates.

5.3 DID model

Furthermore, I employ news consistency as a proxy for media censorship by regulators, as it can more accurately capture nationalist sentiments in countries with varying political systems. To mitigate potential biases introduced by external factors when measuring nationalism through news consistency and to examine the impact of censorship on international conflicts, I utilise a natural experiment involving the 2019 US-China Trade War in my empirical analysis. To determine the change in the measures of national relations associated with the trade war due to news consistency in each entity, I use a Difference-in-Difference (DID) model. DID model is a statistical technique used to assess the causal impact of an intervention or treatment by comparing changes in the outcome variable of interest between a treatment group and a control group before and after the intervention. The DID model estimates the difference in the average change in the outcome variable between the treatment and control groups before and after the intervention, and this difference is considered as the treatment effect. This methodology is widely employed in economics, public health, and social sciences to evaluate the impact of policies, programs, or interventions [30–32].

The DID model I construct with entity and time effects is as follows:

$$y_{it} = \theta_0 + \theta_1 D(T - war)_t \times c_{it} + \omega c_{it} + \zeta_i + \varepsilon_{it}, \qquad (8)$$

where y_{it} is the dependent variable for the measures of US-China relations. In my empirical study, I choose the lagged media sentiment as the proxy for interrelations. $D(T - war)_t$ is the dummy indicator of the US-China Trade War period and ζ_i is source fixed effects. I allow the entity-fixed effects to capture the heterogeneity of each entity measured. I assume that if omitted variables change over time but are constant across all newspapers, the daily media sentiment and consistency indicators should absorb the effect. Therefore, I do not use time-fixed effects. ω is the estimated coefficient to explain how news consistency affects my measures of national relations. ω can be interpreted as the effect of the respective nationalism of the US and China on their bilateral relations without significant authorities' control during a period without conflict. θ_0 is a constant term.

The coefficient θ_1 is my parameter of interest to explain the change in news consistency after the US-China Trade War. The dummy variable $D(T - war)_t$ is used to distinguish between the treatment and control groups. The coefficient θ_1 on the treatment indicator variable represents the difference in the outcome variable between the treatment group ($D(T - war)_t = 1$) and the control group ($D(T - war)_t = 0$) at the baseline (the trade war). The coefficient θ_1 in Eq. (3) captures the marginal effect of daily news consistency of newspaper i at time t on my measures of national relations, indicating the extent to which the common voice of mass media after the trade war affects the outcome variable. A positive (negative) value of θ_1 suggests an increase (decrease) in the effect of news consistency on the outcome variable. Any significant difference in θ_1 between the United States and China can be primarily attributed to the level of media freedom. Notably, China has a system of official news censorship, which ensures that most media reports align with, or at least do not contradict, the government's position. Therefore, news consistency in China during a conflict is likely to reflect to some extent the stance of the central government on particular issues. In contrast, countries with higher media freedom may experience less government interference and therefore may not adequately represent the official position. Consequently, the value of θ_1 is mainly attributable to the change in nationalism caused by the trade war. The sign of θ_1 reveals whether the news consistency during the trade war had a positive (improving international relations) or negative effect (harming international relations) on the outcome variable.

6. Empirical results

This section presents the main results of this chapter. For my main results, I focus on news articles published during the terms of President Barack Obama and President Donald Trump from 2009 to 2019. Using observations from the two presidential terms reflects the stability of US political and diplomatic strategies towards China before and after the trade war. Unlike the competition between Republicans and Democrats in the US, members of the United Front in China fully obey the CCP and must accept its "leadership role" as a condition of their continued existence. Therefore, China's policy towards the US depends mainly on the position of the CCP, which is relatively consistent. Therefore, the relations between China and the US can be effectively measured during the study period.

6.1 LDA results

The main objective of the topic model is to identify news topics, and the results are presented in **Figures 6** and 7 for the US and Chinese news corpora, respectively. The heatmaps in these figures represent the top 12 terms for each topic in the respective corpus after preprocessing, and the darker shades indicate a higher probability that the terms can explain the topics in their respective rows. Although each topic is not assigned a name during the unsupervised learning process of the topic model, the terms grouped in each topic provide a natural annotation. For instance, Topic 2 in the US news corpus, as shown in **Figure 6**, corresponds to "trade war", which is Topic 19 in the Chinese news corpus presented in **Figure** 7.

When the popularity index is closer to 1 (0), it indicates that the topic receives more (less) attention from the masses. In the US news corpus, the most popular topics are "trade negotiation" (1) and "trade war" (2) and the least popular topics are "Wanda Group" (842) and "Dalai Lama" (843). For the Chinese news corpus, the most popular topics are "family" (1), "Belt and Road Initiative" (14) and "trade war" (19), and the least popular topics are those related to the "US Congress" (531) and "US political parties" (534). The distribution of news topics from US and Chinese newspapers in my data set is depicted in two histograms in **Figure 8a** and **b**. It can be observed that news topics with higher popularity indices tend to have more reports, which aligns with my expectations. Popular topics are more likely to be reported by the media, and thus the popularity index can accurately reflect readers' interest in the topics.

6.2 Trade war effects

To accurately estimate the significance of the β coefficient, I use cluster-robust standard errors because within-individual media error correlations may still remain.

Table 4 presents the estimates of the news sentiment Σ_{it} (see Section 4.2 for more details) associated with all of the news topics in my corpora. For Chinese news articles related to the US, we can see a significant increase in sentiment after the trade war, leading to an overall increase in sentiment among Chinese newspapers towards the

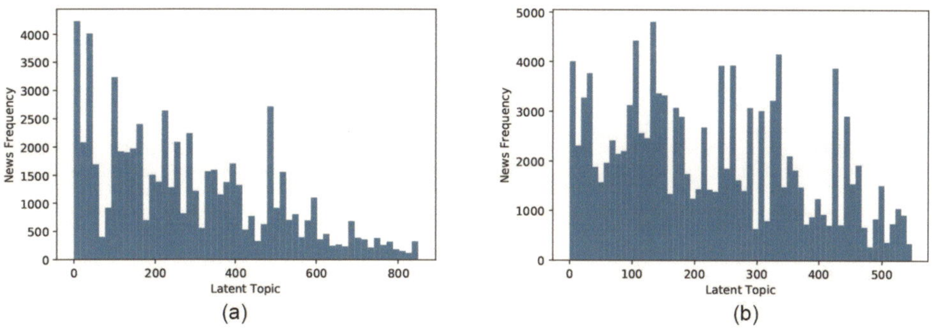

Figure 8.
Distribution of News Topics. Note: *These histograms figure out the distribution of latent topics generated by the Topic Model over my news corpus. There are 64,026 relevant news reports in the US from 1980 to 2019 and 123,549 records in China from 1999 to 2019. The topic number starting from zero is in the sequence of its popularity index (from highest to lowest). (a) The United States, (b) China.*

US. For all articles published between 2009 and 2019, we can see a significant decrease in news sentiment in US newspapers and a significant increase in Chinese newspapers. The decline in news sentiment in US newspapers towards China is in line with my expectations, as the rise of nationalism during conflicts between two nations often triggers negative emotions against the other country. In contrast, the rise of positive descriptions of the US in Chinese newspapers deserves our attention. After controlling the macro factors and the topic-specific factor, I find the main difference between US and Chinese newspapers is media freedom. Therefore, I attribute the positive effects of the trade war on sentiment in Chinese newspapers to the government's efforts to weaken the negative impact of the trade conflict on international relations with the US.

Among my control variables, we observe that the degree of political uncertainty is significantly negatively correlated with news sentiment in US and Chinese newspapers (except for non-negative Chinese news articles), whereas financial market performance is generally positively correlated with news sentiment. On exception is negative Chinese news articles, which are negatively correlated with market performance. This can be explained as follows. When the financial market does not perform well in China, dissatisfaction with the domestic financial market prompts citizens to aspire to a free capital market like the US. Conversely, when the financial market is overheated, a higher sense of national superiority makes people more confident in the country's economy and increases disdain for its main competitors. For the topic-specific control, I find that both popular and unpopular topics are significantly related to news sentiment (except for negative Chinese news articles). One interesting observation is that for popular news articles, US newspapers are more likely to negatively describe China-related issues. In contrast, Chinese media tend to positively present news related to the US in both popular and unpopular topics. I attribute this phenomenon to newspapers meeting readers' tastes. If the constant effect can fully control the specific reporting bias of each newspaper, we can infer that US readers in general are more interested in negative Chinese news, whereas Chinese readers are more interested in positive US news. Furthermore, if the topic is not interesting, for example, religious topics, the media in both countries do not report it negatively.

To clarify the economic importance of my estimated coefficients, I use the term "t-war effect" to represent the value of trade war coefficient $\hat{\beta}_1$ as a percentage of the average measures of US-China relations before the US-China trade war and use stars

Main Regressors	US (1)	China (2)
D(T-war)	−0.037***	0.054***
	[.002]	[.009]
CPU Index	−0.000085***	0.000013**
	[.000]	[.000]
Market Return$_{t-1}$	0.35***	0.40***
	[.090]	[.195]
D(Popular)	−0.0051***	0.011
	[.000]	[.008]
D(Unpopular)	−0.0017	0.0020
	[.001]	[.003]
Constant	0.43***	−0.054***
	[.001]	[.011]
Observations	42,144	81,861
Source FE	Yes	Yes
Pre-war avg	0.42	−0.034
T-war effect	−8.9***	−158.6***

Notes. This table reports the estimating results on newspaper overall sentiments from the US and Chinese sources. Dependent variables include the average sentiment of total articles. Cluster-robust standard errors in parentheses. Coefficients are labelled based on significance (p < .1, ** p < .05, ***p < .01). The T-war effect indicates the estimated coefficient on a dummy D(T-war) as a percentage of the average score of the daily average newspaper sentiment in the region before May 2, 2016. I report the same significant star labels as the level of estimated coefficient on D(T-war) within the column.*

Table 4.
Results of the effects of the trade war on national relations.

to indicate the statistical significance of the estimated effect. This indicator aims to measure the degree and direction of the trade war's influence on media sentiment. If the value is positive, the trade war amplifies the media sentiment. If the value is higher, the amplified effect will be more significant. For instance, the estimated t-war effect in **Table 4**, Column (1), is $\frac{-0.037}{0.42} \times 100 \approx -8.9$. As the sign of the corresponding pre-war average is negative, the t-war effect represents the average effect of the trade war, contributing to an 8.9% opposite change in the overall sentiment in US newspapers about China after the trade war period. Comparing the empirical results in the US and China, the t-war effect of China sources dominates that of the US sources, indicating that the trade war event had a more significant impact on descriptions of average sentiment in Chinese media during the sampling period. Moreover, although the US and China media changed their attitude during the trade war, this conflict significantly reduced the number of negative narratives in Chinese newspapers about the US.

Table 5 shows the results of the Trade War's effects on bilateral imports. Unsurprisingly, I observe that the trade war had a significant negative effect on US and Chinese bilateral imports. We can see that the trade war had a significantly stronger negative effects on US demand for Chinese products and services. If we only focus on the economic consequences of the trade war, it seems that US customers suffered

Main regressors	US (1)	China (2)
D(T-war)	−0.20***	−0.045***
	[.003]	[.001]
CPU index	−0.00025***	−0.000032***
	[.000]	[.000]
Market return_t-1	0.25***	0.049***
	[.057]	[.012]
Population	3.79***	−0.39***
	[.102]	[.019]
D(Dispute)	−0.030***	−0.017***
	[.002]	[.001]
Constant	−0.75***	0.69***
	[.033]	[.025]
Observations	36,428	70,113
Source FE	No	No
News region	US	China
Pre-war avg	0.41	0.14
T-war effect	−50.2***	−33.0***

Notes. This table reports the estimating results on imports in the US (China) from China (the US). Dependent variable is the total value of all types of imported goods (in trillions of dollars). Standard errors in parentheses. Coefficients are labelled based on significance ($p < .1$, ** $p < .05$, *** $p < .01$). The T-war effect indicates the estimated coefficient on a dummy D(T-war) as a percentage of the average value of the dependent variables before May 2, 2016. I report the same significant star labels as the level of estimated coefficient on D(T-war) within the column.*

Table 5.
Results for the effects of the trade war on imports.

greater losses, which is consistent with Waugh [21]. However, as the trade war was launched by President Trump, I attribute the stronger negative effect on US imports of Chinese products and services to a trade-off with the objective of gaining more bargaining power during the negotiation process.

Based on the content of the Phase One trade deal, my results support the argument that the US has stronger bargaining power during the negotiations. Based on the estimated coefficients of the macro controls, I observe a significantly negative effect of political uncertainty and WTO dispute cases on imports and a positive effect of the domestic population and financial market performance. One interesting observation is that in the US, the population determines the level of imports of Chinese products and services, while this effect is not significant in China. I attribute this result to the fact that China has a relatively complete industrial system, so there is less demand for importing daily consumer products from the US.

6.3 News consistency effects

In **Table 6**, I report how the popularity of news topics affects sentiment consistency, with all features in my model being significant. One interesting result is that popular news topics tended to be more consistent after the trade war in China as β_1

Main regressors	US (1)	China (2)
D(T − war) × Popularity	−0.021**	0.259***
	[0.009]	[0.020]
Popularity	0.038***	−0.070***
	[0.008]	[0.018]
D (Political Bias)	−0.003***	—
	[0.001]	—
Chinese Ratio	−0.125***	—
	[0.034]	—
Constant	0.929***	0.768***
	[0.001]	[0.001]
Observations	42,145	85,617

Notes. Standard errors in parentheses. Coefficients are labelled based on significance ($p < .1$, ** $p < .05$, *** $p < .01$).*

Table 6.
News consistency results.

from Eq. 4 in Column (2) is significantly positive, whereas in the US the result is the opposite. Moreover, from the constant terms, we see the newspapers about China topics have higher media consistency in the US. We may also find similar conclusions since the average media consistency in the US (**Figure 5a**) is higher than China's (**Figure 5b**), and the average media consistency after 2016 (yellow dot line) slightly increased in China whereas it decreased in the US. The news censorship system in China may provide a plausible explanation for these differences. Although all published news articles are subject to censorship in China, the degree of censorship varies depending on the objectives of the authorities. It is reasonable to believe that if a news topic is more popular, media regulators expend more energy on reviewing its content and in removing any "unwanted" description. However, it also increases its influence on content manipulation because it attracts more readers' attention. Therefore, using the trade war, we observe different strategies adopted by the Chinese government in response to hot news topics in different periods. Newspapers in the US provide a mixed sample of a news system without central management. As most media in the US are profit oriented, to a large extent, the more popular the news, the more it needs to respond to the tastes and ideologies of subscribers. However, different groups of American readers had different opinions on whether to launch or continue the trade war. Therefore without central control by the central government, US newspapers should have low news consistency if other factors are the same as the Chinese media. However, as we see from the constant terms in **Table 6** and the average consistency in **Figure 5**, the US's coefficient is significantly higher than China's. One explanation is that albeit the Chinese regulators could censor the media topics and contents, they still have some limitations in their ability to control the sentiments of news about the US. Except for a few media with official backgrounds, regulators might not fully control most of the media's reporting sentiments during non-special periods. Another explanation is that although the US media is less manipulated by the regulators, the US media has a more unified sentiment towards China in terms of market or profit-driven purposes. For example, if the US public generally believes that the US should decouple or should maintain good relations with China, then market forces will naturally drive media consistency in coverage of China. It can be seen from the empirical results that the market has a stronger impact on media consistency than regulators.

Table 7 reports the DID estimates of the lagged media sentiment in Eq. (8) based on my data set after 2009. An important result is the positive coefficient θ_1 in Column (2), indicating that the public media in China systematically reported US news more positively during the trade war. As the Chinese government has a strong influence on the national media and some Chinese media abroad, we can reasonably conclude that Chinese regulators tried to soften domestic anti-US sentiment amid the Sino-US trade conflict, which can be seen as an effort by Chinese regulators to repair Sino-US relations. In opposite, the negative coefficient θ_1 in Column (1) denotes the public media in the US spontaneously reported China news more negatively during the trade war.

To clearly explain the role of news consistency after the trade war in the US-China relations, I create an indicator called "unity effect" to estimate coefficient $\hat{\theta}_1$ as a percentage of $\hat{\theta}_1$ multiplied by a parameter κ divided by the average of the observations of the dependent variable y_{it} before the trade war. κ is a threshold value used to determine whether a news article should be recognised as consistent with that of other media. I choose $\kappa = 0.9$ for US newspapers and $\kappa = 0.8$ for Chinese newspapers as these values are close to the mean of the frequency distribution of the documents in **Figure 8**. Therefore, $\kappa\hat{\theta}_1$ indicates consistent news articles, with the estimated measures of US-China relations changing by at least this amount due to the trade war. For instance, the unity effect in **Table 7**, Column (2), is $0.8 \times \frac{0.072}{-0.034} \times 100 \approx -167.9$. I add stars next to the unity effect values estimated based on the significance level of $\hat{\theta}_1$ that I use to generate the indicator. Based on the unity effect in **Table 7**, I conclude that the impact of Chinese and US nationalism on international relations has undergone opposite changes during the trade war. Before the trade war, nationalism in China, represented by media consistency, could more significantly reflect Sino-US relations changes. The negative ω in Column (2) denotes Chinese nationalism weakening

Main regressors	US (1)	China (2)
D(T − war) × Consistency	−0.043***	0.072***
	[.002]	[.013]
Consistency	0.033***	−0.063***
	[.012]	[.013]
Constant	0.39***	0.0058
	[0.011]	[0.005]
Observations	42,144	81,866
Source FE	Yes	Yes
Pre-war avg	0.42	−0.034
Unity effect	−9.3***	−167.9***

Notes. Cluster-robust standard errors in parentheses. Coefficients are labelled based on significance ($p < .1$, ** $p < .05$, *** $p < .01$).*

Table 7.
DID Results for the Sentiment Measures.

China's relations with the US. During the trade war, the θ_1 in Column (2) became significantly positive and contributed to China's relations with the US. Note the Chinese regulators partly censor the media consistency as a proxy of nationalism. Therefore, one explanation of nationalism's positive international relations effect is regulators' willing of the improving Sino-US relations. In opposite, I do not observe a similar result in the US. From the US perspective, the escalating trade conflict between the US and China has naturally soured relations with China.

6.4 Placebo-controlled study

To address concerns regarding the impact of pre-existing factors on the results presented in **Table 7**, a placebo-controlled study is conducted to assess the news sentiment as described in Eq. (8) in US and Chinese newspapers prior to the 2019 US-China trade war. The same DID model employed in Section 6.3 is utilised, with θ_1 representing the difference in the outcome variable between the treatment group $(D(T-war)_t = 1)$ and the control group $(D(T-war)_t = 0)$ at the baseline. However, the placebo test differs from the original analysis in that the baseline is set to December 5, 2004 for US newspapers and December 5, 2007 for Chinese newspapers, which were periods without significant bilateral conflicts based on the Sino-US relations. This placebo study covers the period from the September 11 attacks to the time before Barack Obama's Asian rebalance policy. If the previous analysis of the impact of the trade war on measures of US-China relations is accurate, the unity effect observed in the placebo test should be insignificant or slightly significant. The results of the placebo test are reported in **Table B.1** in Appendix B, and no significant effects are observed except for the constant term in Column (1).

Overall, my placebo-controlled study suggests that the results presented in Section 6.3 are robust to potential confounding factors. Specifically, I find that the difference in the unity effect between the treatment and control groups is primarily driven by the trade war event rather than pre-existing heterogeneous factors.

7. Robustness

In **Table C.1** in Appendix C, I present the results of my robustness tests for the main results in Section 6. Each table reports the coefficients of the t-war effect or the unity effect and their significance is based on the coefficients used in the calculations. The estimated sign of these effects and their level of significance in each robustness test are very close to the main results discussed in Section 6. Using a period of more intense conflict as the trade war event, I observe that the estimated values of the t-war effect and the unity effect tend to be higher, indicating that nationalism and government intervention in the media tend to be stronger as international conflicts escalate.

In Section 4.3, I present the LDA topic model I use to determine whether a news article should be classified as popular. As previously mentioned, K is an important parameter that we need to provide to the machine using the LDA topic model to accurately and efficiently identify the latent topics of each text document. **Figure A.1** in Appendix A presents the validation perplexity for different numbers of topics for the US and Chinese news corpora. To test the robustness of my results, I modify the news popularity dummy by increasing the number of latent topics K in the topic model to reclassify a topic for each news article, then compute a new popularity index for each new topic. I choose $K = 900$ for the US news corpus and $K = 600$ for the Chinese news corpus in this test by enhancing the model's ability to identify latent topics. **Table C.1** in Appendix C reports the results of this robustness test for the main results in Section 6.2. Under the new setting, the estimated t-war effect is as significant as in **Table 4**. Based on these results, I confirm my main conclusion on the effects of the trade war on US-China bilateral relations in the public media.

8. Conclusions

In my study, I have found evidence supporting my hypothesis that nationalism and international disputes play crucial roles in shaping bilateral relations between countries. Specifically, I have used the trade war effect and unity effect to capture the impact of nationalism and news censorship on US-China relations. My results show that the trade war has resulted in a reduction in civil and economic relations between the parties in conflict. However, news censorship can help mitigate or even conceal the nationalism generated by media sentiment, thus aiding in the repair of bilateral relations.

Based on my empirical findings, I have drawn two significant conclusions. First, excessive nationalism can hinder progress in negotiations during international conflicts, particularly if they are asymmetric, and may not always serve a government's interest. Second, in instances where one party has weaker bargaining power, suppressing unfavourable nationalism among its citizens may facilitate negotiation agreements that ultimately increase the total utility of the nation, even if some view the agreement as unfair. Overall, this chapter proposes a methodology based on textual analysis and topic modelling to quantify subjective nationalism and international relations. Such techniques are gaining popularity in the social sciences, with experts predicting a significant impact on various research fields and the management of political and economic risks.

As Shiller argues, textual analysis is poised to become a solid field in economics, owing to the increasing volume of big data generated by social media and interdisciplinary developments in fields such as psychology, neuroscience, and artificial intelligence [33]. Thus, machine learning and text analysis methodologies hold great promise for driving revolutionary results in social science and other research domains.

Acknowledgements

I would like to express my gratitude for all of the invaluable help that I have received during my research. First, I would like to express my sincere gratitude to Professor Xiangdong Wei and Professor Cheng-ze Simon Fan. They convincingly guided and inspired me with promising research ideas and directions. Next, I would like to express my sincere gratitude to Professor Shang-jin Wei, who supported my research project and provided me with many creative academic ideas and professional advice. In addition, I thank Shenzhen Polytechnic Research Fund (6023310005S) and the editor Erick Giovani Sperandio Nascimento.

Thanks

To my parents, Hua and Xing.

Abbreviations

AI	Artificial Intelligence
CCP	Chinese Communist Party
CPU	Chinese Political Uncertainty
DID	Difference-in-Difference
LDA	Latent Dirichlet Allocation
NLP	Natural Language Processing

A. Topic selection

In the main analysis, I select 850 topics for the US news corpus and 550 topics for the Chinese news corpus. To determine if our choice of model is appropriate, I test the goodness of fit of the topic model with different numbers of latent topics K. First, I randomly choose a 20% subset of documents as our testing data and use the remaining documents as our training data to fit the LDA model for K between 50 and 1000. Then I calculate the validation perplexity of the model with the testing data as a measure of the goodness of fit. Perplexity indicates how well the model describes a set of articles, a lower value suggesting a better fit. **Figure A.1** shows the validation perplexity of the testing data for the US and Chinese news corpora. For the US news corpus, the validation perplexity stays flat after $K = 850$, and for the Chinese news corpus, the value increases after $K = 550$. Based on these observations, I choose the model for our main analysis that can best fit the data.

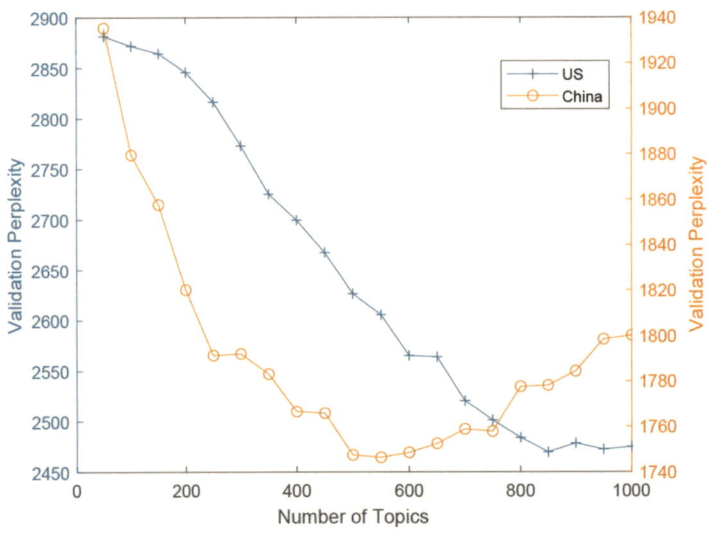

Figure A.1.
Validation perplexity of testing data for different topics. Note: *This figure indicates the validation perplexity of 20% random selected documents for testing. Typically when the number of topics increase, the accuracy of the model enhances, and the model with the lowest perplexity is generally considered the "best". These data show that the goodness-of-fit of the topic model reaches the apex around $K = 850$ for English documents and $K = 550$ for Chinese documents.*

B. Placebo estimates on news sentiment

Main regressors	US (1)	China (2)
D(T − war) × Consistency	−0.0041	0.016
	[.004]	[.016]
Consistency	−0.0022	−0.048
	[.015]	[.094]
Constant	0.44***	−0.027
	[0.015]	[0.065]
Observations	6484	8722
Source FE	Yes	Yes
Pre-war avg	0.44	−0.064
Unity effect	−0.8	−20.5

*Notes. This table replicate the estimation procedure for the **Table 7** under the placebo-controlled study setting. I set the observation window of the US newspaper from January 2002 to December 2006 and choose December 5, 2004 as the date of a placebo event. For Chinese newspaper, I set the data from January 2005 to December 2010 and set December 5, 2007 as the placebo date. Cluster-robust standard errors in parentheses. Coefficients are labelled based on significance (* $p < .1$, ** $p < .05$, ***$p < .01$). The unity effect for the US region reports the estimated coefficient $\hat{\theta}_1$ (see details in Section 6.3) times 0.9 (approximate mean value to distinguish inconsistent and consistent news in **Figure 5a**, and for Chinese newspaper the value is 0.8 corresponds to **Figure 5b**), and divided by pre-placebo average media sentiment. The significant stars of the unity effect correspond to the significance level of the parameter $\hat{\theta}_1$ that I estimated.*

Table B.1.
Placebo estimates for news sentiment.

C. Robustness for nationalism

Main regressors	US (1)	China (2)
D(T-war)	−0.037***	0.054***
	[.002]	[.009]
CPU Index	−0.000084***	0.000014***
	[.000]	[.000]
$MarketReturn_{t-1}$	0.30***	0.39**
	[.098]	[.164]
D(Popular)	−0.0078***	0.0065**
	[.001]	[.003]
D(Unpopular)	0.0010	0.0018
	[.002]	[.008]
Constant	0.43***	−0.053***
	[.002]	[.009]

Main regressors	US (1)	China (2)
Observations	42,133	81,964
Source FE	Yes	Yes
Pre-war avg	0.42	−0.034
T-war effect	−8.9***	−159.5***

Notes. This table denotes the robustness test for news articles with the top 35% most popular topics. The corresponding baseline results are presented in ***Table 4****. In this test I modify the number of latent topic in the Topic model from 850 to 900 for the US news and from 550 to 600 for the Chinese news as plotted in* ***Figure A.1****. Cluster-robust standard errors in parentheses. Coefficients are labelled based on significance (*$^{*}p < .1$*,* $^{**}p < .05$*,* $^{***}p < .01$*).*

Table C.1.
Robustness results for the effects of the trade war.

Author details

Michael D. Wang
Shenzhen Polytechnic, Shenzhen, China

*Address all correspondence to: wangdongmichael@gmail.com

References

[1] McCombs ME, Shaw DL. The agenda-setting function of mass media. Public Opinion Quarterly. 1972;**36**(2): 176-187

[2] McCombs M, Shaw D. The Agenda-setting Function of the Press. Oxford, England: Oxford University Press Inc; 2005. pp. 156-168

[3] Luo Y. The internet and agenda setting in China: The influence of online public opinion on media coverage and government policy. International Journal of Communication. 2014;**8**:24

[4] Chan A. Guiding public opinion through social agenda-setting: China's media policy since the 1990s. Journal of Contemporary China. 2007;**16**(53): 547-559

[5] Mead WR. China is the Real Sick Man of Asia. 2020. Available from: https://www.wsj.com/articles/china-is-the-real-sick-man-of-asia-11580773677

[6] Eriksen TH. Nationalism and the internet. Nations and Nationalism. 2007; **13**(1):1-17

[7] Hyun KD, Kim J, Sun S. News use, nationalism, and internet use motivations as predictors of anti-japanese political actions in China. Asian Journal of Communication. 2014;**24**(6): 589-604

[8] Zhou M, Wang H. Anti-Japanese sentiment among Chinese university students: The influence of contemporary nationalist propaganda. Journal of Current Chinese Affairs. 2017;**46**(1): 167-185

[9] Breton A. The economics of nationalism. Journal of Political Economy. 1964;**72**(4):376-386

[10] Coenders M, Scheepers P. The effect of education on nationalism and ethnic exclusionism: An international comparison. Political Psychology. 2003; **24**(2):313-343

[11] Kuzio T. Nationalism, identity and civil society in Ukraine: Understanding the orange revolution. Communist and Post-Communist Studies. 2010;**43**(3): 285-296

[12] Moffat JE. Nationalism and economic theory. Journal of Political Economy. 1928;**36**(4):417-446

[13] Solt F. Diversionary nationalism: Economic inequality and the formation of national pride. The Journal of Politics. 2011;**73**(3):821-830

[14] Chan I, Li M. New Chinese leadership, new policy in the South China Sea dispute? Journal of Chinese Political Science. 2015;**20**(1):35-50

[15] Fravel MT. Explaining stability in the senkaku (diaoyu) islands dispute. Getting the Triangle Straight: Managing China-Japan-US Relations. 2010;**2010**: 144-164

[16] Storey IJ. Creeping assertiveness: China, the Philippines and the South China Sea dispute. Contemporary Southeast Asia. 1999;**1999**:95-118

[17] Hayes CJH. Essays on Nationalism. Macmillan; 1928

[18] Heilmann K. Does political conflict hurt trade? Evidence from consumer boycotts. Journal of International Economics. 2016;**99**:179-191

[19] Liu T, Woo WT. Understanding the us-China trade war. China Economic Journal. 2018;**11**(3):319-340

[20] Li C, He C, Lin C. Economic impacts of the possible China–us trade war. Emerging Markets Finance and Trade. 2018;**54**(7):1557-1577

[21] Waugh ME. The consumption response to trade shocks: Evidence from the us-china trade war. Journal of International Economics. 2019;**118**:158-173

[22] Berthou A, Jardet C, Siena D, Szczerbowicz U. Quantifying the losses from a global trade war. Banque de France ECO Notepad. 2018;**19**:1-4

[23] Montiel CJ, Salvador AMO, See DC, De Leon MM. Nationalism in local media during international conflict: Text mining domestic news reports of the China–Philippines maritime dispute. Journal of Language and Social Psychology. 2014;**33**(5):445-464

[24] Pandit S, Chattopadhyay S. Coverage of the surgical strike on television news in India: Nationalism, journalistic discourse and India–Pakistan conflict. Journalism Practice. 2018;**12**(2): 162-176

[25] Lorentzen P. China's strategic censorship. American Journal of Political Science. 2014;**58**(2):402-414

[26] King G, Pan J, Roberts ME. How censorship in China allows government criticism but silences collective expression. American Political Science Review. 2013;**107**(2):326-343

[27] Hansen S, McMahon M, Prat A. Transparency and deliberation within the fomc: A computational linguistics approach. The Quarterly Journal of Economics. 2017;**133**(2):801-870

[28] Baker SR, Bloom N, Davis SJ. Measuring economic policy uncertainty. The Quarterly Journal of Economics. 2016;**131**(4):1593-1636

[29] Wang MD, Lou J, Zhang D, Fan CS. Measuring political and economic uncertainty: A supervised computational linguistic approach. SN Business & Economics. 2022;**2**(5):37. DOI: 10.1007/s43546-022-00209-2

[30] Rubin DB. Estimating causal effects of treatments in randomized and nonrandomized studies. Journal of Educational Psychology. 1974;**66**(5):688

[31] Bertrand M, Duflo E, Mullainathan S. How much should we trust differences-in-differences estimates? The Quarterly Journal of Economics. 2004;**119**(1):249-275

[32] Card D, Krueger AB. Minimum Wages and Employment: A Case Study of the Fast Food Industry in New Jersey and Pennsylvania. The American Economic Review. 1994;**84**(4):772-793

[33] Shiller RJ. Narrative economics. American Economic Review. 2017; **107**(4):967-1004

Chapter 3

A Quantitative Analysis of Big Data Analytics Capabilities and Supply Chain Management

Janine Zitianellis

Abstract

With the emergence of Big Data Technologies (BDT) and the growing application of Big Data Analytics (BDA), Supply Chain Management (SCM) researchers increasingly utilize BDA due to the opportunities from BDT and BDA present. Supply Chain (SC) data is inherently complex and results in an environment with high uncertainty, which presents a real challenge for SC decision-makers. This research study aimed to investigate and illustrate the application of BDA within the existing decision-making process. BDT allowed for the extraction and processing of SC data. BDA aided further understanding of SC inefficiencies and delivered valuable, actionable insights by validating the existence of the SC bullwhip phenomenon and its contributing factors. Furthermore, BDA enabled the pragmatic evaluation of linear and nonlinear regression SC relationships by applying machine learning techniques such as Principal Component Analysis (PCA) and multivariable regression analysis. Moreover, applying more sophisticated BDA time series and forecasting techniques such as Sarimax, Tbats, and neural networks improved forecasting accuracy. Ultimately, the improved demand planning and forecast accuracy will reduce SC uncertainty and the effects of the observed SC bullwhip phenomenon, thus creating a competitive advantage for all the members within the SC value chain.

Keywords: big data analytics, supply chain management, bullwhip phenomenon, principal component analysis, regression analysis, demand planning and forecasting

1. Introduction

BDA presents an opportunity for precise and transparent information flow between crucial SC components such as procurement, inventory management, and demand planning and forecasting and encourages SC integration and collaboration to promote overall SC efficiency [1, 2].

SCM ultimately attempts to match varying supply and demand rates in the most cost-efficient manner. As one can imagine, the flow of materials and information through several organizations within the SC network is inherently and increasingly complex due to various SC channels and data nodes driven by factors such as market globalization and SC digitalization initiatives [3]. SC inefficiencies will manifest and contribute to the well-examined bullwhip phenomenon [4].

IntechOpen

This research critically evaluated the relationship between BDA capabilities and SC performance within a single case study. Therefore, the primary focus of this research was to investigate and illustrate the application of BDA within the existing decision-making process to overcome technology constraints and challenges within the SC information flow and obtain valuable insights into the current market environment. This research aimed to understand better and measure the interaction between in-store sales and in-store stockholding as a dynamic market indicator aiding optimal demand planning and forecasting. The following research question addresses specific SC issues, such as reducing operational costs, risks, and the financial impacts associated with demand forecast inefficiencies and missed market opportunities:

"What BDA methods improve SC demand planning and forecasting in SMEs?"

Even though growing research into the application of BDA within SMEs is evident, there are limited case studies that illustrate the application of BDA reflecting measurable business value, such as a reduction in operational cost, risk, and the financial impact associated with demand forecast inefficiencies and missed market opportunities. Thus, this research project aimed to advance practical knowledge within the area of interest, applying various BDA techniques, increasing SC collaboration and communication, and ultimately improving SC efficiency by employing a single organization case study within the SME industry.

2. Big data, analytics, and supply chain management within the SME industry

2.1 Big data and analytics

In 1941, the term "information explosion" was followed by several publications between 1944 and 2000, reflecting on the magnitude and expected growth rate of data and information. It remains unclear as to the true origin of the term "Big data." While many states the term was popularized by computer scientist John R Mashey it is believed that the term was only officially coined in 2005 by Roger Mougalas and the O'Reilly Media group [5], describing big data as large datasets that cannot be processed or consumed through traditional business processes and analytical tools [6]. Gartner, a leading technology research and consulting firm, extends a more recent definition by incorporating critical characteristics of big data and describing big data as: "...high-volume, high-velocity and/or high-variety information assets that demand cost-effective, innovative forms of information processing that enable enhanced insight, decision making, and process automation" [7]. Several studies report on the characteristics of big data, often described as the five Vs of big data or big data dimensions. Namely, these are volume, variety, velocity, value, and veracity [8–10]. Volume and variety describe the data's size, magnitude, and format. At the same time, velocity, value, and veracity summarize how closely the data is processed in real time, the business value the information generates, and the trust in the data quality. There have been several extensions to the big data Vs in recent years, including adding variability. Variability considers the change in data structures and, more importantly, the pace, frequency, and extent to which those data structures change. Understanding the variability of one's data, supported by a high level of data veracity, allows for efficient planning of available resources and processes to ensure minimal disruption to the organization's decision-making process.

Big Data Analytics BD, coupled with analytics, defined as BDA, presents scalable and cost-effective opportunities to process and integrate different data structures, enabling the extraction of valuable insights from high volume and various structured and unstructured data. Kitchin [11] examined the impact of BDA on established methodologies and how big data acts as a disruptive innovation while permitting new and more efficient analytical methods to emerge. Technical skills are vital in extracting information and insights from that data effectively. Those analytical efforts also rely on clearly articulated business objectives and the evaluation of results by subject matter experts.

The insights drawn allow organizations to respond to rapidly developing environments and attain a competitive advantage on tactical and strategic levels [1, 9]. An empirical study by [9] argues that organizations require agility to reconfigure operations to respond to BDA insights delivered and generate actionable insights to obtain optimal business value from BDA initiatives. The author observes that even though organizations within Norway are progressive in adopting information and technology advances, only some organizations observe the complete business benefit of their BDA investment. Surprisingly, the study reports a low 10.8% of companies with more than 4 years of BDA experience, considering a notable increase in BDA trends over the past decade [9]. Findings in a study by [1] suggest organizations still lack an understanding of the required BDA capabilities to enable BDA as a strategic driver. The study emphasizes that the success of BDA initiatives within an organization is dependent on a composite factor of the allocation of time and financial resources, a supportive BDA infrastructure, and leadership commitment to driving a data-driven culture.

2.2 Big data analytics within supply chain management

BDA presents an opportunity for precise and transparent information flow between crucial SC components such as procurement and inventory management, encouraging SC integration and collaboration [1, 2]. Furthermore, the computational efficiency of BDA accommodates the increasing complexity of SC data, driven by factors such as market globalization, increased market competition, and SC digitalization initiatives [3]. Thus, BDA allows for more accurate demand planning and forecasting underpinning SC inventory management decisions to promote overall SC efficiency.

SCM was defined by [12] as the alignment and horizontal integration of organization, supplier, and customer processes that tie customer demand with capital, materials, services, and information. The definition is consistent with the description provided by [13], who described SCM as the interconnection between organizations through processes to satisfy consumer demand. While excess stock appears to be the conventional approach to managing supply and demand rate fluctuations, holding excess stock has several disadvantages, including the increased risk of stock losses through theft or damages. Fisher [14] expresses the importance of implementing SC policies to ensure downstream stock levels are kept at a minimum, driving increased throughput rate and reducing the working capital otherwise engaged in stock.

The early works of Professor Marshal Fisher stressed the critical importance of prompt information sharing, enabled by innovative technologies and methods, allowing SC members the opportunity to adjust lead times and efficiently respond to changes in market demand. Still, demand forecasting necessitates accurate SC data and the sharing of appropriate SC information by all the members within the SC [2, 15].

To fully appreciate the complexity of SC demand planning and forecasting, one must understand the focal point of demand and supply: inventory management. The core commodity within the SC is inventory, alternatively, stock. Slack et al. [13] describe stock "as the accumulation of material resources within a transformation system." Conjointly, inventory planning aims to ensure an optimal inventory level, decreasing the risk and cost of stock-outs and carrying excessive stock to efficiently respond to changes in the customer demand [13].

SC decision-maker supports these objectives by implementing the appropriate inventory management policies and executing replenishment strategies informed by forecasted demand.

The accuracy of the forecasted demand is crucial in support of the following critical decisions:

- The volume decision: What and how many stock items to order?
- The timing decision: When to place the stock order?
- The inventory analysis and control decision: What policies and procedures support the decision-making process?

The efficiency of the inventory decision-making process contributes significantly to the bullwhip effect. The bullwhip describes the consequences of minor interruptions within the demand side of the SC, escalating into significant disruptions, such as amplified demand creating variability in replenishment orders moving up the SC [16, 17].

A rigorous evaluation of the bullwhip effect by [4] examines the bullwhip effect as a clear indication of SC inefficiency. Several studies have postulated a convergence between demand signal processing, lead times, order batching, shortage gaming and rationing, price fluctuations, and behavioral causes as factors causing SC disruptions and contributing significantly to the bullwhip effect directly associate with the three major SC inventory decisions [4], Authors [13, 14] suggest that an abnormal rise in SC costs is evidence of deterioration in several supply chains due to self-serving SC relationships and poor and unnecessary price promotion practices causing SC disruptions, increasing SC uncertainty, and leading to inadequate SC performance.

BDA can facilitate SC management by delivering valuable insights and improving demand forecasting. Studies [1, 3] discussed a notable increase in the literature between 2015 to 2019, particularly research that applied more sophisticated BDA techniques in demand forecasting to improve accuracy. The authors illustrate deployment of more sophisticated machine learning methods, including but not limited to neural networks, regression, ARIMA, Support Vector Machine (SVM), and decision trees. These methods address the drawbacks of conventional time series techniques, such as the high reliance on domain knowledge and the inability to incorporate external factors and compute complex non-linear customer demand behavior and relationships. They are believed to outperform conventional methods [3].

The computational efficiency of BDA accommodates the increasing complexity of SC data, allowing for more accurate forecasting and predictions underpinning SC inventory decisions. Thus, improved forecast accuracy can lessen the effects of the observed SC bullwhip phenomenon, reflecting measurable business value, including but not limited to a reduction in operational cost, risk, and the financial impact associated with demand forecast inefficiencies and missed market opportunities.

2.3 Big data analytics within the SME industry

BDA presents an opportunity for SMEs that do not have the resources to invest in costly systems and data analytics infrastructures to leverage the same technologies and capabilities as their larger counterparts [8]. For example, SMEs can utilize BDA techniques to aid and improve essential SC functions such as procurement, inventory management, and demand planning and forecasting.

However, despite the SME industry being universally acknowledged as a crucial sector, much of the literature focuses on the benefits to larger organizations, demonstrating a gap in understanding BDA within the SME industry. A study by [18] revealed that 32.8% of South African SMEs considered the rapidly changing technological environment a critical challenge, with 28.24% concerned with the high cost associated with information technology. These challenges faced by SMEs influenced the implementation of strategic initiatives to reduce operational costs, increase profits, and create a competitive advantage. Similarly, [19] expressed concerns about the slow adoption and integration of technology and innovation within the SME sector and drew attention to little progress in promoting awareness of BDA opportunities. Thus, much uncertainty still exists about the adoption and relevance of BDA within the SME industry.

3. Research methodology

3.1 Design

The main research objectives are advancing practical knowledge to an identified research problem to reduce SC uncertainty and improve demand planning and forecasting, mainly through applying BDA, a developing field within the SME industry. Due to the complexity surrounding SC processes and data, this research adopted a pragmatic position, and the research premise is built on existing theory and concluded as valid. Subsequently, the research is underpinned by a deductive research approach which involves testing causal relationships of two or more concepts of variables set out in a series of hypotheses within the theory boundaries and conditions [20] further describes the deductive research approach as a well-known approach within social research, deducing and subjecting the hypothesis to empirical examination. In more simplistic terms, deductive research flows from theory to data. The use of a quantitative design inherently increases the possibility of research generalizability.

Deploying a single case study research strategy guided the necessary research actions in a structured and linear fashion and promoted coherence throughout the research project. The research's descriptive and explanatory nature allowed for deeper insights and knowledge into the role of BDA in improving demand planning and forecasting, which desired a longitudinal design to explore the relationship between consumer demand and supply within brick-and-mortar stores.

3.2 Techniques and procedures

The BDA component of this research utilized the CRISP-DM framework [21]. The CRISP-DM is a cross-industry standard process for data mining, developed initially in 1996 to form data mining projects. Today, the methodology is still relevant and successfully applied within the BDA and data science field [22]. With a strong emphasis on business understanding and the constant alignment with business objectives throughout the process, several activities transformed the data into actionable insights to better understand the research problem and support the research objectives. The chosen analysis types are driven by the nature of the research and are closely aligned with the research question, namely **descriptive**, **diagnostic**, and **predictive** analysis. The descriptive nature of the analysis was conducted on a bivariate level to explore any relationship between the measure variables and determine the magnitude or impact of a change in one variable on the other, most often measured by the change in the respective mean values. Understanding the relationship is achievable through correlation analysis and contingency tables. However, it is crucial to note that the bi-variate analysis measures the relationship, draws attention to the effect of change on one another, and by no means implies causality [20].

Throughout this research analysis, the widely used Pearson correlation method was adopted to understand the strength and significance of the linear relationship relative to the unit of analysis [23]. A drawback of the Pearson correlation is the requirement for normally distributed data, as it is a parametric method. Thus, to further support initial correlation findings, non-parametric methods such as Kendall's tau_b and Spearman's rank (rho) were deployed to achieve correlation synthesis. Furthermore, adjustments to the interpretation of the correlation scores were necessary, considering the nature of slow-moving non-food products with sparse data [24]. Thus, any variable reflecting a Pearson and Spearman's (rho) correlation score greater than 0,2 was considered a fair association.

A series of machine learning techniques namely, multivariable linear regression and regression trees, and variable reduction technique, namely principal component analysis (PCA) were explored to pursue a further in-depth analysis and address this research's diagnostic nature. As Jim [23] states, regression analysis applies various statistical processes to understand the nature of the relationship between dependent and independent variables. Therefore, regression analysis is used to further analyze the relationship between supplier-retailer and retailer-consumer demand variance and the bullwhip phenomenon. The regression equation estimates the relationship and identifies factors of importance that influence the bullwhip measure. In addition, multivariable regression allows for more than one independent variable in the model, whereby each variable has an additive contribution toward the change in the dependent variable [25]. However, a known limitation of regression methods is the sensitivity to highly correlated independent variables, known as multicollinearity. Consequently, multicollinearity introduces data redundancy, impacts the statistical significance of variables, and reduces the precision of estimated model coefficients [26]. A solution was assayed for exploiting multicollinearity using techniques such as principal component analysis (PCA). According to James et al. [25], PCA compares common variation between variables and generates a series of uncorrelated, linear combinations or indexes known as components that collectively explain the most significant proportion of variance within the dataset. Furthermore, it can reduce known regression limitations, such as overfitting, if the PCA assumptions available in

Appendix E hold true [27]. Finally, the predictive nature of the research deployed various time series and forecasting techniques performed within RStudio statistical software. Analysis results were visualized within the Tableau tool stack.

3.3 Ethical considerations

Data protection and ethical considerations within a B2B context mainly focus on compliance with the retailer's data storage and usage policies and procedures. It was of utmost importance that the B2B portal data and in-store observations were processed securely, and usage was access controlled, treating any sensitive retailer data as "personal" data. Furthermore, the same principles and consent were considered for the B2B data in compliance with the General Data Protection Regulation (GDPR) framework [28]. The extracted data were stored in a CSV format on a secure designated Google Drive for collection and analysis. The necessary procedures and activities were carried out to anonymize the data.

4. Case study: South African SME supplier of slow-moving consumer goods

The South-African SME, Zeus Africa, competes as a polymer (HDPE) toy ride-on bikes manufacturer within the South African toy industry. Industry expert [29] reviews the sector as a challenging yet promising market. Thus, harnessing innovative technological capabilities to improve demand planning and forecasting accuracy and support effective decision-making is critical for the organization's survival and competitive advantage in current market conditions. A key consideration is the information flow and capability of the chosen organization to integrate their B2B data and observational store data into the existing decision-making process to derive valuable consumer insights and reduce supply chain uncertainty.

4.1 Data description

The case study involves identifying the problem and its relevant objectives and hypotheses. The critical sources employed to generate the required information are the following:

4.1.1 Primary internal quantitative financial data

The data extracted from the organization's primary financial system consist of sales orders spanning over 3 years, from August 2018 to October 2021. The data include transaction-level daily sales orders of seven unique SKUs delivered to a single retailer.

4.1.2 Internal secondary quantitative B2B data

The data extracted from the organization's B2B portal consist of weekly aggregated sales and stock holding data for brick-and-mortar stores, spanning from August 2018 to October 2021.

4.1.3 Integrated data containing supplier sales orders and store sales and inventory data

The analysis was supported by an additional integrated dataset, namely "B2B store replenishment," containing supplier sales orders and sales and inventory data at the store level. This dataset was aggregated on a store and SKU ID level.

The data was needed to understand the relationship between **supplier-retailer** and **retailer-consumer demand** and store stock levels. Sales levels are assumed to be sensitive to in-store stockholding, and the unit of analysis for this study was the relationship between these components within the retail brick-and-mortar stores, considering the following demand scenarios:

- Supplier-retail sales orders to adequately meet retailer-consumer demand.
- Retailer-consumer demand when stock is available.
- Retailer-consumer demand when there is insufficient stock or no stock is available.

These units contribute to an understanding of historical patterns and trends of consumer demand relative to demand planning over time. Moreover, understanding SC uncertainty by measuring the influence of stock supply on consumer demand and the overall impact on upstream SC demand. In addition, exploring the variance between supplier-retailer and retailer-consumer demand substantiates the existence of the bullwhip phenomenon.

4.2 Descriptive analysis and discussion

The descriptive analysis aims to identify critical SKUs through the application of bivariate techniques and establish if sufficient evidence within supplier-retailer and retailer-consumer demand variance suggests the bullwhip effect.

4.2.1 Evidence of the bullwhip phenomenon

The analysis considered the SD values of the supplier sales order quantity, the store stock replenishment, and sales quantity demand as the unit of analysis to validate the existence of the bullwhip effect. A higher SD sales order quantity value is observed than store stock replenishment and sales quantity demand. Thus, demonstrating a higher variance in supplier-retailer demand than retailer-consumer demand. Subsequently, the bullwhip ratio was derived by adapting the metric used by [4]. The bullwhip at a supplier-retailer demand was determined by measuring the variance ratio between supplier sales order quantity and the store stock replenishment, see Equation (1).

Zeus Africa supplier bullwhip ratio:

$$Supplierbullwhipratio = sku\frac{std.dev(SalesOrderQuantity)}{std.dev(Replenishment(StockIn))} \tag{1}$$

The bullwhip at a retailer-consumer demand was determined by measuring the variance ratio between store stock replenishment and the sales quantity demand, see Equation (2).

Zeus Africa retailer bullwhip ratio:

$$Retailerbullwhipratio = sku\frac{std.dev(Replenishment\ (StockIn))}{std.dev(SalesQtyDemand)} \tag{2}$$

Interpretation of the bullwhip ratio is relatively simple. A bullwhip ratio of one indicates an equal variance between demand and supply. Therefore, no upstream SC demand amplification is evident. Conversely, a bullwhip ratio of less than one indicates that supply is less variable than demand. In conclusion, a bullwhip ratio greater than one indicates amplified demand variability. **Figure 1** represents the bullwhip ratio obtained for supplier-retailer demand and the retailer-consumer demand for each SKU.

Inspection of supplier-retailer demand in **Figure 1** revealed sufficient evidence of demand amplification for all SKUs except SKU ID BZ030, arguably due to the short time in the market, indicating that the consumer demand variance translated into amplified sales order quantity variance. The descriptive analysis yielded sufficient evidence to validate the bullwhip effect leading to a further in-depth analysis of the factors driving the bullwhip phenomenon discussed in the upcoming Section 4.3.

4.3 Diagnostic analysis and discussion

The objective of the diagnostic analysis within the research context was to validate the driving factors contributing to the observed bullwhip phenomenon attributed to the supplier-retailer and retailer-consumer demand variance through regression analysis techniques. In addition, the diagnostic and predictive analysis techniques employed a stratified sampling method to support the research study's validity and reliability, and the sample outcome is available in Appendix A.

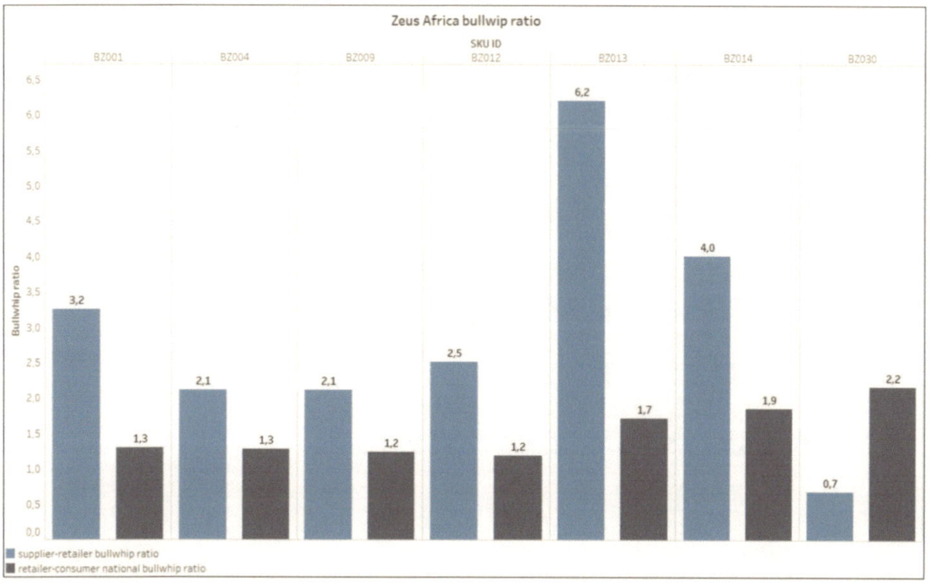

Figure 1.
Zeus Africa supplier and retailer bullwhip ratio.

4.3.1 Regression analysis

The preceding Section 2.2 highlights demand signal processing, lead times, order batching, behavioral causes, shortage gaming and rationing, and price fluctuations as major contributing factors to the bullwhip phenomenon. The analysis considered the calculated store bullwhip ratio presented in Equation (3) as the dependent variable and variables associated with store sales quantity demand, stock levels, and stock replenishment variance as the independent variables.

Zeus Africa retailer internal (store) bullwhip ratio:

$$Storebullwhipratio = store, sku \frac{std.dev(Replenishment(StockIn))}{std.dev(SalesQtyDemand)} \tag{3}$$

The regression analysis further analyzes the relationship between supplier-retailer and retailer-consumer demand variance and the bullwhip phenomenon. The regression equation estimates the relationship and identifies factors of importance that influence the bullwhip measure. Discussing the application of regression analysis relevant to this research project focuses on the business benefit by highlighting the impact and the degree of change in the bullwhip measure due to demand variability and SC uncertainty.

An exploratory analysis of the stores' bullwhip ratio indicates that more than 50% of the store and SKUs recorded notably high bullwhip ratios implying significant variances between the store stock replenishment and sales quantity demand. Furthermore, **Figure 2** revealed a much higher mean store bullwhip ratio than the national bullwhip ratio. Thus, a national bullwhip ratio potentially conceals stores recording excessive bullwhip ratios and subsequently overlooks SC inefficiencies.

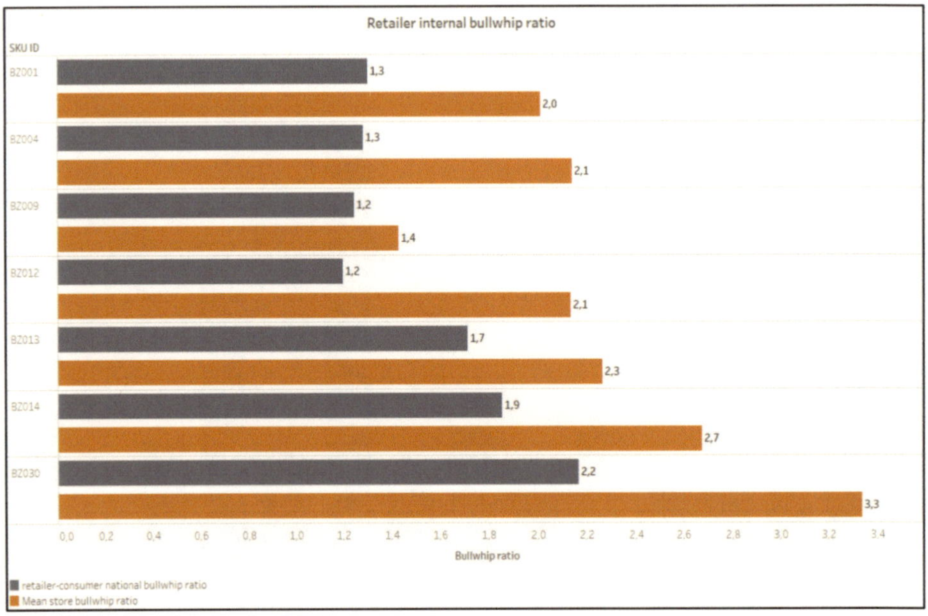

Figure 2.
Zeus Africa retailer store bullwhip ratio.

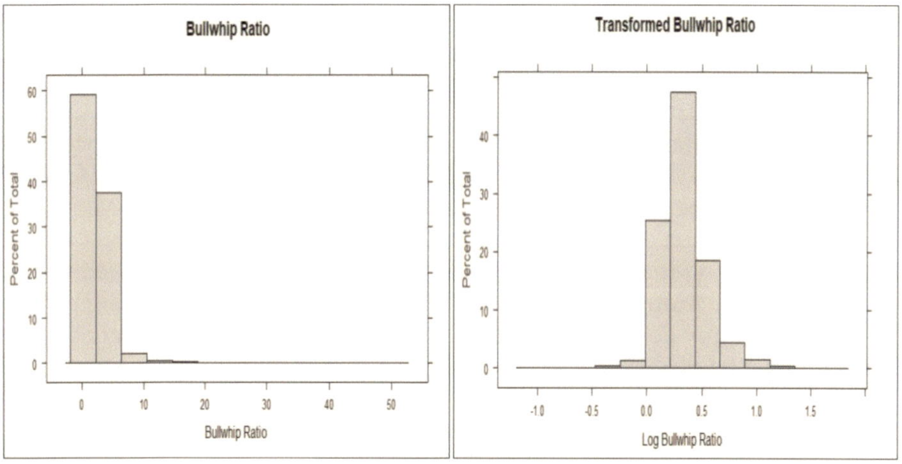

Figure 3.
Zeus Africa logarithmic transformation (log10) of the bullwhip ratio.

Determining the strength and significance of the linear relationship between the bullwhip ratio and the variables mapped to the contributing factors was applying the correlation methods set out in Section 3.3. However, to satisfy the Pearson correlation assumption of normality, a logarithmic transformation (log10) of the bullwhip ratio was necessary to reduce skewness and achieve a near-normal distribution [30], represented in **Figure 3**.

The bullwhip ratio correlation coefficients and covariance interpretation are limited to the variables satisfying the condition available in Appendix B. Specifically, the analysis reveals that high stock levels over long periods are associated with an increase in the store log-transformed bullwhip ratio versus higher stock turn ratios. Reducing the overall time that SKUs are stocked in a store is associated with a decrease in the store's log-transformed bullwhip ratio.

Furthermore, the correlation analysis highlighted the existence of multicollinearity within the independent variables, available in Appendix C. Multicollinearity introduces data redundancy, impacts the statistical significance of variables, and reduces the precision of estimated model coefficients [26]. However, principal component analysis (PCA) exploits multicollinearity, compares common variations between variables, and generates a series of linear combinations or indexes known as components [27].

4.3.2 PCA analysis

According to [25], PCA produces uncorrelated components comprised of the most optimal linear combination of variables and collectively explains the most significant proportion of variance within the dataset. Furthermore, it can reduce known regression limitations such as overfitting, if the PCA assumptions hold true. Each principal component (PC) represents a proportion of variance explained (PVE). The first PC often carries the most significant PVE value, and the objective is to include as few PCs as possible in the regression model while explaining the most cumulative variance. Author [25] highlights that while there is no single approach to determine the optimal number of PCs, an intuitive inspection of the scree plots detailing each PC PVE and the cumulative PVE would highlight the optimal number of PCs.

A review of the scree plots available in Appendix E and relevant to this research project yielded cumulatively that four PCs explain 79% of the variance within the dataset. In addition, the model accuracy was assessed by evaluating the root mean square of residuals (RMSR). The PCA model yielded an RMSR value of 0.056, which is on the cusp of the acceptable threshold value of 0.05 [31].

An overview of each component generated and the associated variable loading is provided. The loading represents the variable coefficient for each generated component, indicating the strength of the association with the component. Strong positive variable loadings provide substantial confirmation that the underlying variables and encoded PCs can be explicitly associated with the contributing factors to the bullwhip phenomenon.

Following is an overview of each component generated and the associated variable loading. The loading represents the variable coefficient to each generated component, thus indicating the strength of the association with the component, presented in **Table 1**.

Variable loadings	PCA (PC1) - Price, Promotion fluctuations	PCA (PC2) - Order batching	PCA (PC3) - Demand signal processing	PCA (PC4) - Lead time
Total demand (sales quantity)	0.88			
Total stock replenishment	0.90			
Total promo indicator	0.78			
Rate of sale (ros)	0.91			
Mean supply (stock quantity balance)	0.79			
Total store count excess stock ind		0.77		
Total month end indicator		0.97		
Weeks total trading		0.97		
Total store count lows ind			0.77	
Store count lows ind ratio			0.76	
Store stock turn ratio			0.70	
Std. store replenishment				0.82
Mean weeks of stock				0.60
Total lead time				0.87
Distance from dc			0.48	
SS loadings	4.16	3	2.58	2.08
Proportion variance	0.28	0.20	0.17	0.14
Cumulative variance	0.28	0.48	0.65	0.79

Table 1.
Zeus Africa PCA variable loadings.

5. Components explain 79% of the variance

5.1 Regression analysis training and testing sets

This allows for further in-depth analysis of the store's bullwhip ratio through the application of regression techniques, which involves creating training and testing datasets to initiate the analysis. By stratifying on SKU ID and dividing the "B2B store replenishment" dataset into training and testing datasets according to a 70/30 ratio split available in Appendix D, whereby the training data train the various algorithms. The testing data aids in an unbiased evaluation of the model's accuracy and predictive power [25]. The fair representation of SKU ID between the training and testing sets was verified to ensure consistency.

5.1.1 Multivariable regression analysis

Multivariable linear regression is a parametric assessment method that makes certain assumptions about underlying data for analysis [32]. **Table 2** represents a summary of the model assumptions and their outcomes.

While all conditions were met, the autocorrelation in the residuals was a source of uncertainty. The non-randomness within the error terms indicates an underlying pattern within the store bullwhip ratio, and these factors are missing from the current data. Alternatively, the model requires adjustment. An adjustment to a non-linear polynomial relationship to the third degree was made. Polynomial relationships describe a curvilinear relationship that displays an increase in the dependent variable with each increase in the independent variable until it reaches a point where subsequent increases result in a decrease. With an adjusted R square value of 0.32, RMSE value of 0.72, and SI value of 0.50, the model did not yield any more success in comparison and was subsequently discarded.

5.1.2 Regression model coefficients and interpretations

The model coefficient describes the expected impact on the dependent variable and the relationship between the dependent and independent variables. **Figure 4** illustrates the relationship between the log-transformed bullwhip measure and each PC at a store and SKU level, underpinned by the regression coefficient output and the relevant business interpretation detailed in Appendix E. Inspection of **Figure 4**

Assumption	Condition satisfied
Assumption 1: The regression model is linear in parameters.	True
Assumption 2: The mean of residuals is zero or close to zero.	True
Assumption 3: Homoscedasticity of residuals or equal variance.	True
Assumption 4: No autocorrelation of residuals.	False
Assumption 5: The number of observations must be greater than the number of Xs.	True
Assumption 6: No perfect multicollinearity.	True
Assumption 7: Normality of residuals.	True

Table 2.
Bullwhip regression model assumptions.

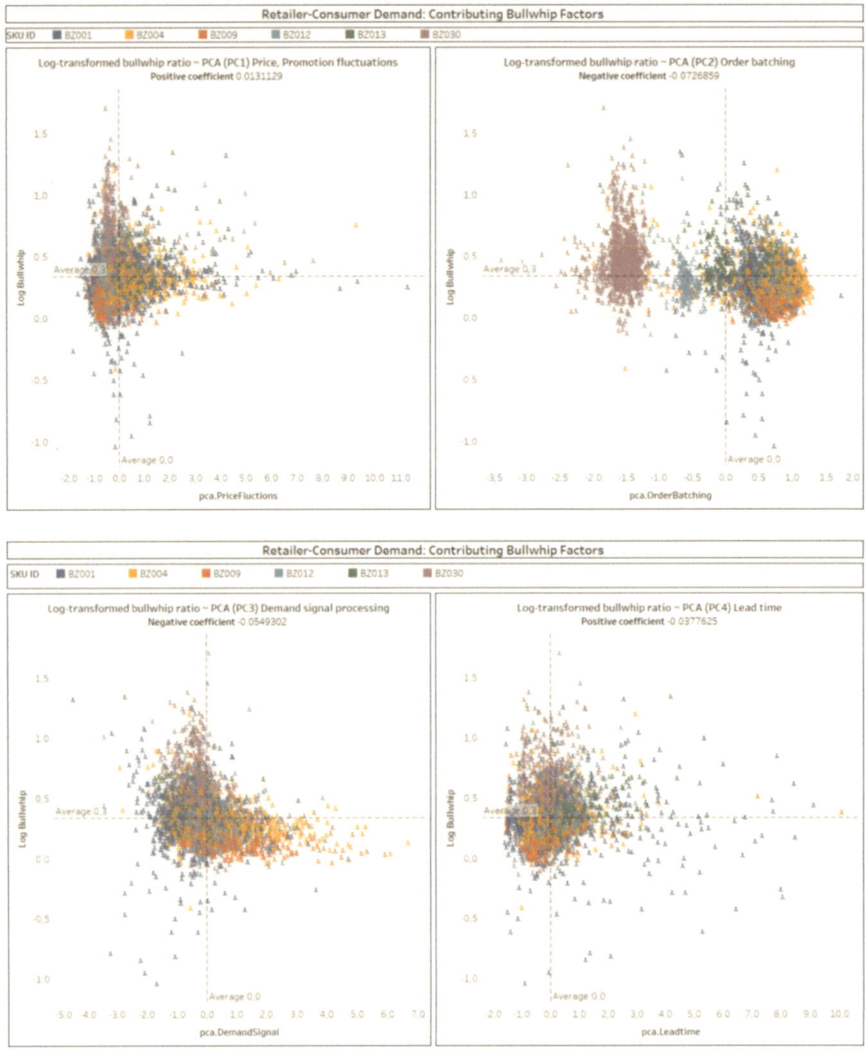

Figure 4.
Zeus Africa log-transformed bullwhip ratio and PCs relationship.

revealed that SKU ID BZ001 is more prone to outliers across all PCs. Furthermore, SKU ID BZ030 is visibly isolated when reviewing the log-transformed bullwhip ratio and PCA (PC2) order batching relationship. Keeping in mind the variables associated with PCA (PC2), namely "total trading weeks" and "total store count excess stock," a possible explanation for the results is the short period that SKU ID BZ030 has been in the market, which aligns with the correlation analysis findings of the bullwhip ratio tends to decrease over time.

5.1.3 Regression model evaluation and accuracy

The final model was measured and assessed at a significance level of 0.05, denoted as $\alpha = 0.05$. Appendix F details each assessment measure, the applicable hypothesis test, and the business-relevant interpretation. The model was applied to the test data and the complete dataset to determine how well the model predicts or generalizes to new data, using the computed RMSE and SI as performance indicators for each dataset compared. The test data reported a lower RMSE of 0.07, indicating no overfitting, and reporting a lower SI of 0.05, which is well within the accepted threshold value of 1. Applying the model to the complete dataset, a slightly increased RMSE value of 0.18 was observed. Notwithstanding, the observed SI value of 0.54 remains within the threshold. Thus, accepting the accuracy of the predictions and concluding the model could generalize well to new data [25]. While not all model assumptions were satisfied, the adjusted R Squared within the context of this research project at 0.31, albeit low, is acceptable considering the complexity of these interlinked bullwhip attributes, the model prediction performance measures, namely SE, RMSE, and SI values, are considered satisfactory and the overall model at P-value <0.05 is significant.

5.2 Predictive analysis and discussion

The objective of the predictive analysis is to predict the likelihood of future events. The primary focus was on improving demand planning and forecast accuracy. Within the context of this research project is the application of BDA incorporating dynamic market demand signals sourced from the retailer B2B data into the Zeus Africa decision-making process.

5.2.1 Time series analysis for demand planning and forecasting

Time series analysis involves analyzing historical data before forecasting. Time series methodologies are classed and considered to be the mining of complex data types yet described as a sequence of ordered events expressed numerically, such as customer demand recorded at equal time intervals [33]. Forecasting customer demand involves transforming the time component into an independent variable and estimating future demand based on observed historical demand and current demand signals.

In a recent review of predictive BDA for supply chain forecasting, authors [3] highlight a growing trend and increase in the application of BDA techniques such as but not limited to Neural Networks, regression, Arima, Support vector machine (SVM), and decision trees within the area of SC demand forecasting. The application of these methods addresses the drawbacks of conventional time series techniques, such as the high reliance on domain knowledge and the inability to incorporate external factors and compute complex non-linear customer demand behavior and relationships.

Seeking evidence of improved SC demand forecasting through BDA techniques, the research employed and grouped five well-established time series methods [34]. The time series techniques were grouped into two categories a) conventional time series techniques, namely Holt-Winters and Arima, and b) BDT-enabled time series techniques, namely Sarimax, Tbats, and Neural Networks. These techniques are briefly described in Appendix B. The forecast accuracy of each of the minimal viable time series models employed was benchmarked using several measures described in Appendix C, followed by a business-relevant interpretation.

5.2.2 Time series models and forecasts

The key variables identified to analyze and forecast supplier-retailer demand was the dependent variable, namely sales order quantity, and the independent variable, namely the partition period representing the "time" element of sales orders delivered to the retailer DC during the analysis period. Furthermore, periods of non-delivery were interpreted as missing values in the analysis. Missing values can potentially introduce model bias and decrease model performance. Several time series models can accommodate missing values, namely Arima and Neural Networks models. However, missing values are problematic for time series regression and Tbats techniques. Consequently, it necessitated value interpolation, imputing missing values [35]. Furthermore, a cubic spline interpolation was employed to accommodate the non-linear relationship between supplier-retailer demand, retailer-consumer demand, and stock supply [36].

Additional independent variables were identified and derived from the retailer B2B data, namely (a) the total mean estimated lead time from placing a stock order and receiving thereof at the retailer brick and mortar store, (b) promotional indicator, (c) national stock supply at the point in time, (d) national rate of sale (ROS), and (e) the absolute number (count) of stores reflecting no stock, low stock or excess stock available at the point in time. These variables were incorporated to improve demand forecast accuracy.

5.2.3 Time series model and forecasting evaluation

The following measures assessed the forecast accuracy and benchmarked each of the minimal viable time series models employed, followed by the business-relevant interpretation. **Table 3** defines the selected model evaluation and assessment measure.

5.2.4 Time series training and testing sets

The standard approach for validating time series model performance is selecting observations into the relevant training and testing data sets, and commonly the testing, also referred to as the validation set, will consist of the most recent data observations. Within the context of this research, the test data set contained observations for the most recent 13 weeks from 11 July 2021 to 3 October 2021, and the training data set included all available observations before 11 July 2021 for each SKU ID.

Measure	Definition
Mean absolute error (MAE)	Represents the average magnitude of errors and is expressed in units of the dependent variable. The absolute fit of the model to the data can be measured by how closely the predicted values align with the actual values.
Mean absolute percentage error (MAPE)	Represents the forecast errors as a % of the actual observed value.
Forecast Errors Root mean squared error (RMSE)	Represents the squared absolute fit of the model to the data, measuring how close the actual values are to the predicted values. The RMSE value is expressed in units of the dependent variable and translates to the standard deviation of the unexplained variance.
Scatter index (SI)	A measure of determining if the RMSE is acceptable, a value of <1 is deemed acceptable.

Table 3.
Time series model evaluation criteria.

5.2.5 Time series model and forecasting for SKU ID BZ001

Each time series method recorded SI values of <1 and within the acceptable threshold. However, the BDA techniques recorded a significantly lower mean MAPE (error rate) of 33.5 than the conventional techniques' mean MAPE (error rate). In addition, the BDA techniques reflect higher forecast accuracy, as evident from lower mean MAE (1172) and mean RMSE (1663) values compared to the mean MAE (1148) and mean RMSE (1882) of the conventional methods. The overall results indicate that both Sarimax and Tbats models are considered the best-performing models, with low MAPE (error rate) values of 27.8 and 23.0, respectively. Accompanied by low MEA values of 975 and 955, respectively, and RMSE values of 1315 and 1466.

5.2.6 Time series model and forecasting for SKU ID BZ004

Each series method recorded SI values of <1 and within the acceptable threshold. However, the BDA techniques recorded significantly lower mean MAPE (error rate) values of 37.5 than the mean MAPE (error rate) of 47.2 for the conventional techniques. In addition, the BDA techniques reflect higher forecast accuracy, as evident from lower mean MAE (648) and mean RMSE (719) values compared to the mean MAE (928) and mean RMSE (1120) of the conventional methods. The overall results indicate that Sarimax is the best-performing model with the lowest MAPE (error rate) value of 23.7 and the lowest MEA value of 300, and RMSE value of 370.

5.2.7 Time series model and forecasting for SKU ID BZ009

Each series method recorded SI values of <1 and within the acceptable threshold. The BDA techniques significantly improved the MAPE (error rate) value of 30.1 for the conventional techniques to the MAPE (error rate) value of 22.0. In addition, the BDA techniques reflect higher forecast accuracy, as evident from lower mean MAE (163) and mean RMSE (196) values compared to the mean MAE (236) and mean RMSE (308) of the conventional methods. The results indicate that the Sarimax model is considered the best-performing model, with a low MAPE (error rate) value of 15.3. Accompanied by a low MEA value of 103 and an RMSE value of 119.

5.2.8 Time series model and forecasting for SKU ID BZ012

Each series method recorded SI values of <1 and within the acceptable threshold. The overall BDA techniques delivered counterintuitive results due to the poor performance of the Sarimax model, recording a high MAPE (error rate) value of 72.3 and a high MEA value of 933, and RMSE value of 1247. However, the overall results indicate similar performance of the Arima, TBATS, and Neural Networks models, reporting closely aligned MAPE, MAE, and RMSE values.

5.2.9 Time series model and forecasting for SKU ID BZ013

Each time series method recorded SI values of <1 and within the acceptable threshold. Even though the Neural Networks model was unsuitable, the remaining

BDA techniques significantly improved the MAPE (error rate) value of 54.9 for the conventional techniques to the MAPE (error rate) value of 19.0. In addition, the BDA techniques reflect higher forecast accuracy as evidenced by lower mean MAE (553) and mean RMSE (685) values compared to the mean MAE (1604) and mean RMSE (1770) of the conventional methods. The results indicate that the Sarimax model is considered the best-performing model, with a low MAPE (error rate) value of 12.8. Accompanied by a low MEA value of 400 and RMSE value of 540.

5.2.10 Time series model and forecasting for SKU ID BZ030

Each time series method recorded SI values of <1 and within the acceptable threshold. The limited trend or seasonality components rendered the Holt-Winters model unsuitable. However, the conventional Arima model outperformed the BDA techniques, as evident from the low MAPE (error rate) value of 3.5 compared to the MAPE (error rate) value of 14.2 for BDA techniques. In addition, the Arima model reflected a higher forecast accuracy, as evident from lower mean MAE (258) and RMSE (322) values compared to the mean MAE (1023) and RMSE (1340) of the BDA techniques. The overall results indicate that the Arima and Neural Networks models are considered the best-performing models, with closely aligned MAPE, MAE, and RMSE values.

5.2.11 Time series model and forecasting summary

Analysis results show that BDA techniques outperform conventional methods when there is sufficient data available that reflects high seasonal trends and fluctuations. In contrast, conventional techniques, mainly the Arima model, were better suited for SKUs with limited historical or low seasonal data. **Table 4** presents each SKU's optimal time series and forecasting techniques.

6. Conclusions and managerial implications

The research premise considers that the efficient management of an organization across the various operational sub-areas will lead to a sustainable competitive advantage within the area of interest of this research, namely SCM [37].

SKU ID/Optimal method	Conventional Time series and forecasting techniques	BDA Time series and forecasting techniques
BZ001		X
BZ004		X
BZ009		X
BZ012	X	
BZ013		X
BZ030	X	

Table 4.
Time series model and forecasting summary.

The computational efficiency of BDA accommodates the increasing complexity of SC data and assists in managing market challenges. Zeus Africa and similar SMEs would gain from investing in open-source BD technologies and relevant BDA skills and techniques, harnessing the capabilities of these innovative technologies, and allowing for more accurate forecasting and predictions underpinning SC inventory decisions. A compressive BDA framework consisting of comprehensively understanding SKU importance and value by integrating descriptive analysis techniques, namely the ABC inventory and bivariate analysis, will inform and enable the adjustment of resources and processes in line with actual consumer demand at the lowest possible cost without compromising consumer satisfaction levels.

Moreover, incorporating external data, such as their B2B data, and integrating key SC performance measures, such as the bullwhip ratio, into their decision-making process will highlight operational inefficiencies and challenges, enabling an informed and data-driven approach and improving sales order quantity forecasts. However, the author recognizes the limitation of a more appropriate bullwhip ratio relative to accommodate slow-moving products that need further refinements and presents avenues for future research.

While the financial impact of compounded demand planning and forecasting inefficiency because of unaddressed distorted consumer demand and store stock supply was not established and presents avenues for future research, improved forecast accuracy can lessen the effects of the observed SC bullwhip phenomenon, reflecting measurable business value, including but not limited to a reduction in operational cost, risk, and the financial impact associated with demand forecast inefficiencies and missed market opportunities. Therefore, applying various BDA techniques results in improved SC efficiencies and thus contributes to the research premise.

Conflict of interest

The authors declare no conflict of interest.

Thanks

I extend my gratitude to Zeus Africa for granting me the opportunity to enter into the supply chain world and for the resources and support to undertake this research study.

A. Sampling

Recognizing factors influencing the required sample size such as (a) the number of independent variables, (b) missing values, (c) high variance observed in the dependent variables [38], a minimum representative sample size for each SKU at supplier-retailer and retailer-consumer demand level was determined and employed a stratified sampling method to support the research study's validity and reliability.

The required sample size for each SKU ID was inadequate for this analysis, given a 0.05 error rate that is relevant to supplier-retailer demand. Notwithstanding, it was acceptable at a 0.1 error rate.

SKU ID/Measure	Deliveries			Sample size provided (n)	The sample size required (n) at (ε)	
	variance	std. dev	std. error	observation	0.05 error	0.1 error
BZ001	0.1	0.4	0.4	70	191	48
BZ004	0.2	0.4	0.4	121	239	60
BZ009	0.1	0.3	0.3	81	169	43
BZ012	0.2	0.5	0.5	73	362	91
BZ013	0.0	0.0	0.0	32	0	0
BZ014	0.2	0.4	0.4	11	252	63
BZ030	0.0	0.0	0.0	11	0	0

The required sample size, given a 0.05 error rate relevant to the retailer-consumer demand, was adequate.

SKU ID/Measure	Deliveries			Sample size provided (n)	The sample size required (n) at (ε)	
	variance	std. dev	std. error	observation	0.05 error	0.1 error
BZ001	5	2.2	1.9	127.340	7719	1930
BZ004	5.2	2.3	2.1	129.689	7790	1993
BZ009	1.1	1	0.9	88.809	1686	422
BZ012	1.7	1.3	1.1	99.859	2607	652
BZ013	2	1.4	1.2	55.158	3127	782
BZ014	4.3	2.1	1.7	18.167	6581	1646
BZ030	1.4	1.2	0.3	30.466	2152	538

B. Interpretation of the log-transformed bullwhip ratio correlation coefficients and covariance

Variable 1	Variable 2	Pearson correlation coefficient	Kendall's tau_b	Spearman's rho	Interpretation
Store Count Excess Stock Ratio	Log-transformed bullwhip ratio	0.25	0.22	0.32	Positive relationship: An increase in the time increments relative to the total period of the store holding excess stock is associated with an increase in the log-transformed bullwhip ratio.
Est. Weeks of Stock (Supply)	Log-transformed bullwhip ratio	0.2	0.23	0.33	Positive relationship: An increase in the number of weeks of stock available is associated with an increase in the log-transformed bullwhip ratio.
Mean Supply (Stock Quantity Balance)	Log-transformed bullwhip ratio	0.21	0.24	0.34	Positive relationship: An increase in the mean stock on hand is associated with an increase in the log-transformed bullwhip ratio.
Weeks	Log-transformed bullwhip ratio	−0.4	−0.26	−0.36	Negative relationship: An increase in the number of weeks the SKUs are ranged in-store is associated with a decrease in the log-transformed bullwhip ratio.
Store stock turn ratio	Log-transformed bullwhip ratio	−0.47	−0.44	−0.6	Negative relationship: An increase in the store stock turn ratio is associated with a decrease in the log-transformed bullwhip ratio.

C. Bullwhip correlation analysis and interpretations

1 - Kendall's tau_b
2 - Spearman's rho
3 - Pearson Correlation

CT ID	Variable	median_store_replenishment	mean_store_replenishment	sd_store_replenishment	skew_store_replenishment	sd_demand	sd_StockIn	bullwhip	demand	avg_weekly_ros	avg_weekly_supply	avg_WeeksOfStock	OOS_IND_ratio	LOWS_IND_ratio	EXC_STOCK_IND_ratio	Distance	promo_ind_ratio	stock_turn_ratio
1	median_store_replenishment	1,0	0,5	0,3	0,0	-0,1	-0,1	0,0	0,0	-0,1	-0,1	0,0	-0,1	0,0	0,0	0,0	0,0	0,1
1	mean_store_replenishment			0,8	0,1	-0,2	-0,2	0,1	0,0	-0,1	0,0	0,3	-0,2	-0,2	0,2	-0,1	-0,1	0,0
1	sd_store_replenishment				0,3	-0,1	-0,1	0,1	0,1	-0,1	0,1	0,3	-0,2	-0,1	0,2	-0,1	0,0	0,0
1	skew_store_replenishment					0,2	0,1	-0,2	0,5	0,4	0,2	0,1	-0,3	0,3	0,1	-0,1	0,1	0,4
1	sd_demand						0,5	0,0	0,6	0,8	0,4	-0,1	-0,2	0,5	0,0	0,1	0,2	0,3
1	sd_StockIn							0,3	0,3	0,5	0,5	0,0	-0,1	0,2	0,1	0,1	0,2	0,0
1	bullwhip								-0,1	-0,1	0,2	0,3	0,0	-0,1	0,2	0,0	0,0	-0,4
1	demand									0,8	0,3	0,0	-0,4	0,5	0,0	0,0	0,2	0,6
1	avg_weekly_ros										0,4	0,0	-0,3	0,5	0,1	0,0	0,2	0,4
1	avg_weekly_supply											0,4	-0,4	0,1	0,4	-0,2	0,1	-0,1
1	avg_WeeksOfStock												-0,4	-0,2	0,6	-0,2	0,0	-0,3
1	OOS_IND_ratio													-0,1	-0,5	0,2	-0,1	-0,1
1	LOWS_IND_ratio														-0,2	0,1	0,1	0,5
1	EXC_STOCK_IND_ratio															-0,2	0,0	-0,3
1	Distance																0,0	0,1
1	promo_ind_ratio																	0,1
1	stock_turn_ratio																	1,0
2	median_store_replenishment	1,0	0,6	0,3	-0,1	-0,1	-0,2	-0,1	0,0	0,0	-0,2	0,1	-0,1	0,0	0,0	0,0	-0,1	0,2
2	mean_store_replenishment			0,9	0,2	-0,2	-0,3	0,1	0,0	-0,1	0,0	0,4	-0,2	-0,2	0,3	-0,2	-0,1	0,0
2	sd_store_replenishment				0,4	-0,2	-0,2	0,1	0,1	-0,1	0,1	0,5	-0,2	-0,2	0,3	-0,2	0,0	0,0
2	skew_store_replenishment					0,4	0,1	-0,2	0,7	0,5	0,3	0,1	-0,5	0,4	0,1	-0,1	0,1	0,6
2	sd_demand						0,7	-0,1	0,7	0,9	0,5	-0,2	-0,2	0,6	0,0	0,1	0,2	0,4
2	sd_StockIn							0,5	0,4	0,6	0,7	0,0	-0,2	0,3	0,2	0,1	0,2	0,0
2	bullwhip								-0,2	-0,1	0,3	0,4	0,0	-0,2	0,3	0,0	0,0	-0,5
2	demand									0,9	0,5	0,0	-0,5	0,6	0,0	0,0	0,2	0,7
2	avg_weekly_ros										0,6	0,0	-0,4	0,6	0,1	0,0	0,2	0,6
2	avg_weekly_supply											0,6	-0,5	0,1	0,6	-0,2	0,2	-0,2
2	avg_WeeksOfStock												-0,5	-0,3	0,8	-0,4	0,0	-0,4
2	OOS_IND_ratio													-0,2	-0,7	0,2	-0,1	-0,2
2	LOWS_IND_ratio														-0,3	0,2	0,1	0,6
2	EXC_STOCK_IND_ratio															-0,3	0,0	-0,4
2	Distance																0,0	0,2
2	promo_ind_ratio																	0,1
2	stock_turn_ratio																	1,0
3	median_store_replenishment	1,0	0,9	0,4	-0,2	-0,1	-0,1	0,0	-0,1	-0,1	0,1	0,2	0,0	-0,1	0,0	-0,1	0,0	-0,1
3	mean_store_replenishment			0,7	-0,1	-0,2	-0,1	0,0	-0,1	-0,2	0,1	0,3	-0,1	-0,1	0,1	-0,1	0,0	-0,1
3	sd_store_replenishment				0,2	-0,2	-0,1	0,0	-0,1	-0,1	0,1	0,3	-0,2	-0,2	0,2	-0,2	0,0	-0,1
3	skew_store_replenishment					0,3	0,1	-0,2	0,5	0,4	0,2	0,0	-0,5	0,2	0,1	-0,1	0,1	0,4
3	sd_demand						0,7	-0,1	0,7	0,9	0,5	-0,1	-0,2	0,5	0,0	0,1	0,5	0,3
3	sd_StockIn							0,3	0,5	0,7	0,6	0,1	-0,1	0,3	0,1	0,1	0,3	0,0
3	bullwhip								-0,1	-0,1	0,2	0,3	0,1	-0,1	0,2	0,0	0,0	-0,3
3	demand									0,9	0,5	-0,1	-0,4	0,4	0,0	0,0	0,4	0,5
3	avg_weekly_ros										0,6	-0,1	-0,3	0,5	0,0	0,1	0,5	0,3
3	avg_weekly_supply											0,5	-0,4	0,0	0,4	-0,2	0,3	-0,2
3	avg_WeeksOfStock												-0,3	-0,2	0,5	-0,3	0,0	-0,3
3	OOS_IND_ratio													-0,1	-0,7	0,2	-0,1	-0,1
3	LOWS_IND_ratio														-0,2	0,2	0,1	0,5
3	EXC_STOCK_IND_ratio															-0,3	0,0	-0,5
3	Distance																0,0	0,2
3	promo_ind_ratio																	0,1
3	stock_turn_ratio																	1,0

D. Zeus Africa multivariable regression training and testing data

SKU ID	BZ001	BZ004	BZ009	BZ012	BZ013	BZ030	Total
Training Freq and %	444	559	342	488	199	641	2673
	16.6%	20.9%	12.8%	18.3%	7.4%	24.0%	100.0%
Testing Freq and %	191	239	146	209	85	275	1145
	16.7%	20.9%	12.8%	18.3%	7.4%	24.0%	100.0%

E. PCA model assumptions

Assumption 1: Sphericity *or existence of collinearity between the variables.*

Bartlett's Test for Sphericity is used to determine whether the intercorrelation matrix comes from a non-linear population. If this is not the case, PCA may not be appropriate, as it relies on constructing a linear combination of the variables. The hypothesis being tested by Bartlett's Test is stated as follows:

Ho: No collinearity between the variables exist.

Ha: Collinearity between the variables exist.

Application of Bartlett's Test at a significance level of 0.05, denoted as $\alpha = 0.05$, yielded a p-value of < 0.038, thus rejecting the null hypothesis and accepting the alternative hypothesis of sufficient evidence that collinearity between the variables exist.

Assumption 2: Sample adequacy.

The KMO (Kaiser-Meyer-Olkin) test is applied to determine if the data is suitable for dimension reduction techniques such as factor analysis or PCA. The following threshold guides interpretation of the KMO statistic, and KMO values can be interpreted as follows (Stephanie, 2016):

KMO values between:

- to 0.49 is considered unacceptable.
- 0.50 to 0.59 is considered miserable.
- 0.60 to 0.69 is considered mediocre.

In summary, KMO values greater than 0.79 indicate an adequate data sample. KMO values less than 0.5 suggest an inadequate data sample and require remedial action. Applying the KMO (Kaiser-Meyer-Olkin) test yielded an overall KMO score of 0.79, concluding that the data sample is adequate. Furthermore, variables recording a KMO score of less than 0.50 were excluded from the PCA.

Assumption 3. Positive determinant of the correlation or variance-covariance matrices The determinant value must be positive, implying a positive symmetric covariance matrix. The assumption was tested by applying the R "det" function. The "det" function yielded a value of 2.75601e-09, concluding that the positive determine value satisfies the assumption.

Assumption 4. PCA scree plot - PVE and cumulative PVE.

The figure below highlights a drop in the PVE after the fourth PC. However, these four PCs can explain 79% of the variance within the dataset. In addition, the model accuracy on four PCs yielded a root mean square of residuals (RMSR) value of 0.056, which is on the cusp of the acceptable 0.05 threshold (Rajput, 2018). Loadings are representative of the eigenvalue for the respective principal component. All components reflect acceptable SS loading values of greater than 1.

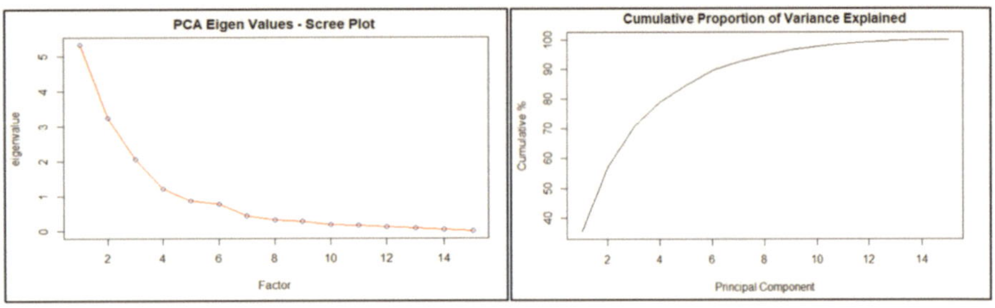

F. Zeus Africa multivariable regression model coefficients and interpretations

Coefficient	Estimate	P-value	Interpretation
Intercept	0.3174756	< 2e-16	The intercept can be interpreted as the average log-transformed bullwhip ratio if all independent variables are set to a value of 0.
SKU ID BZ001	0.3174756	< 2e-16	The estimated value of SKU ID BZ001 is the base value and is equivalent to the intercept. Translating to a mean log-transformed bullwhip ratio of 0.3174756.
SKU ID BZ004	0.0651518	4.58e-08	Stores stocking and selling SKU ID BZ004 will increase the mean log-transformed bullwhip measure by 0.0651518 compared to the base SKU ID BZ001.
SKU ID BZ009	−0.0430543	0.002237	Stores stocking and selling SKU ID BZ009 will decrease the mean log-transformed bullwhip measure by −0.0430543 compared to the base SKU ID BZ001.
SKU ID BZ012	0.0226517	0.062349 ** Significant at $\alpha = 0.01$	Stores stocking and selling SKU ID BZ012 will increase the mean log-transformed bullwhip measure by 0.0226517 compared to the base SKU ID BZ001. While SKU ID did not reflect as significant at an error rate of 0.05, however, accepting an error rate of 0.01 denoted as $\alpha = 0.01$ is considered significant.
SKU ID BZ013	0.0705915	4.50e-06	Stores stocking and selling SKU ID BZ013 will increase the mean log-transformed bullwhip measure by 0.0705915 compared to the base SKU ID BZ001.
SKU ID BZ030	0.0704256	0.000277	Stores stocking and selling SKU ID BZ013 will increase the mean log-transformed bullwhip measure by 0.0704256 compared to the base SKU ID BZ001.
Median weeks store replenishment	−0.0028166	0.003893	For every increase in the median weeks between store replenishment, the mean log-transformed measure will decrease by 0.0028166.
PCA (PC1) - Price, Promotion fluctuations	0.0131129	0.000460	For every unit increase in the PCA price promotion component, the mean log-transformed bullwhip measure will increase by 0.0131129.

Coefficient	Estimate	P-value	Interpretation
PCA (PC2) - Order batching	−0.0726859	< 2e-16	For every unit increase in the PCA order batching component, the mean log-transformed bullwhip measure will decrease by −0.0726859.
PCA (PC3) - Demand signal processing	−0.0549302	< 2e-16	For every unit increase in the PCA demand signal processing component, the mean log-transformed bullwhip measure will decrease by −0.0549302.
PCA (PC4) - Lead time	0.0377625	5.55e-14	For every unit increase in the PCA lead time component, the mean log-transformed bullwhip measure will increase by 0.0377625.

G. Zeus Africa regression model evaluation and interpretations

Measure	Hypothesis/Definition	Result	Interpretation
F critical value and overall p-value	H_o: Intercept only model fits data H_a: Model fits data better than intercept only model	F Stat: 125.1 P-value: < 2.2e-16 With the p-value <0.05 we Fail to reject H_o and accept H_a	We can conclude that the overall model is acceptable, and there is a statistically significant relationship between the store bullwhip ratio and the contributing bullwhip principal components and store demand planning variables identified.
Adjusted R Squared	The independent variables explain the proportion of the dependent variable with adjustment for the number of terms in the model.	Adj. R Square: 0.31	The contributing bullwhip principal components can explain 31% of the variance in the store bullwhip ratio, and store demand planning variables identified.
P-value	H_o: Independent variable does not correlate with the dependent variable. H_a: Independent variable is correlated with the dependent variable.	P-value: < 2.2e-16	Any contributing bullwhip principal components and store demand planning variables with p-value >0.05 were excluded from the model. It is accepted that a statistically significant relationship exists between the remaining bullwhip principal components and store demand planning variables and the store bullwhip ratio.
Standard Error (S)	A measure of goodness of fit is expressed in absolute terms. The following rule of thumb applies to measure the typical distance of the data points from the regression line. The standard error should be smaller than one standard deviation.	Log-transformed S: 0.17 S: 1.5 Log-transformed Std.Dev: 0.21 Std.Dev: 1.6	The standard error of 0.17 is less than one standard deviation of 0.21 and concludes that the model has the required level of precision.
Root mean squared error (RMSE)	Represents the absolute fit of the model to the data, measuring how close the actual values are to the predicted values. RMSE value represents the unit of the dependent variable and translates to the standard deviation of the unexplained variance.	Log-transformed RMSE: 0.17 RMSE: 1.5	The bullwhip ratio ranges from −0.82 to 1.44 units with a standard deviation of 0.21, in context the RMSE value of 0.17 is acceptable and conclude the model is a good fit to the data.

Measure	Hypothesis/Definition	Result	Interpretation
Scatter index (SI)	A measure of determining if RMSE is acceptable, a value of < 1 is deemed acceptable.	SI: 0.39	The scatter index of 0.39 is below the threshold, thus accepting the integrity of the predictions.

H. Time series technique class

Technique	Description	Class
Holt-Winters	An exponential smoothing forecast method. Transforming demand (response variable) into a weighted average of past observation values, whereby current observation values will carry more weight or importance than older observations.	Conventional
Arima (AutoRegressive Integrated Moving Average)	AutoRegressive - Output is regressed on its own lagged observation values. Integrated - The number of times differencing needs to be applied to achieve stationarity. Moving average - using past forecast errors as opposed to past observation values.	Conventional
Sarimax (Seasonal ARIMA + Exogenous variables)	Incorporates all ARIMA components and accommodates exogenous variables – external regressors to improve forecast accuracy.	BDA
Tbats (Trigonometric seasonality Box-Cox transformation ARMA errors Trend Seasonality + Exogenous variables)	An exponential smoothing forecast method. Accommodating for multiple seasonal patterns by the use of a trigonometric function and allowing for exogenous variables - external regressors to improve forecast accuracy.	BDA
Neural Networks	Accommodates non-linear relationships between demand (response variable) and exogenous variables - external regressors.	BDA
SMA Simple Moving Average	Transforming demand (response variable) into an arithmetic average by the number of periods within a given range.	Conventional

I. Time series model measure

Measure	Definition
Mean absolute error (MAE)	Represents average magnitude of errors and expressed in units of the dependent variable. The absolute fit of the model to the data, measuring how close the actual values are to the predicted values.
Mean absolute percentage error (MAPE)	Represents the forecast errors as a % of the actual observed value.
Forecast Errors Root mean squared error (RMSE)	Represents the squared absolute fit of the model to the data, measuring how close the actual values are to the predicted values. The RMSE value is expressed in units of the dependent variable and translates to the standard deviation of the unexplained variance.
Scatter index (SI)	A measure of determining if RMSE is acceptable, a value of <1 is deemed acceptable.

Author details

Janine Zitianellis
Monarch Business School Switzerland, Cape Town, South Africa

*Address all correspondence to: janine.zitianellis@umonarch-email.ch

References

[1] Cetindamar D, Shdifat B, Erfani S. Assessing big data analytics capability and sustainability in supply chains [Internet]. 2020. [cited 2022 Aug 17]. Available from: http://hdl.handle.net/10125/63765

[2] Mafini C, Muposhi A. Predictive analytics for supply chain collaboration, risk management and financial performance in small to medium enterprises. Southern African Business Review. 2017;**21**(1):311-338

[3] Seyedan M, Mafakheri F. Predictive big data analytics for supply chain demand forecasting: methods, applications, and research opportunities. Journal of Big Data. 2020 Jul 25;7(1):53

[4] Disney SM, Lambrecht MR. On replenishment rules, forecasting, and the bullwhip effect in supply chains. now Publishers. 2008. [cited 2022 Jul 4]. [Internet] Available from: https://ofppt.scholarvox.com/catalog/book/10232240?_locale=en

[5] Firican G. The history of big data [Internet]. LightsOnData. 2022. [cited 2022 Aug 8]. Available from: https://www.lightsondata.com/the-history-of-big-data/

[6] Dontha R. Who came up with the name big data? - DataScienceCentral.com. Data Science Central. [Internet]. 2017. [cited 2022 Aug 8]. Available from: https://www.datasciencecentral.com/who-came-up-with-the-name-big-data/

[7] Definition of Big Data - Gartner Information Technology Glossary [Internet]. Gartner. [cited 2022 Aug 8]. Available from: https://www.gartner.com/en/information-technology/glossary/big-data

[8] Iqbal M, Kazmi SHA, Manzoor A, Soomrani AR, Butt SH, Shaikh KA. A study of big data for business growth in SMEs: Opportunities & challenges. 2018

[9] Mikalef P, Krogstie J, Pappas IO, Pavlou P. Exploring the relationship between big data analytics capability and competitive performance: The mediating roles of dynamic and operational capabilities. Information and Management. 2020 Mar 1;**57**(2):103169

[10] Oncioiu I, Bunget OC, Türkeş MC, Căpuşneanu S, Topor DI, Tamaş AS, et al. The impact of big data analytics on company performance in supply chain management. Sustainability. 2019; **11**(18):4864

[11] Kitchin R. Big data, new epistemologies and paradigm shifts. Big Data & Society. 2014;**1**(1):1-12

[12] Krajewski LJ, Malhotra MK, Ritzman LP. Operations Management. Processes and Supply Chains. 11th ed. England: Pearson; 2016

[13] Slack N, Chambers S, Johnston R. Operations Management. 5th ed. United Kingdom: Pitman Publishing; 2007

[14] Fisher ML. What is the right supply chain for your product? [Internet]. 1997. [cited 2021 Jan 1]. Available from: https://www.academia.edu/31156494/What_Is_the_Right_Supply_Chain_for_Your_Product

[15] Mathu KM. The information technology role in supplier-customer information-sharing in the supply chain management of south African small and medium-sized enterprises. South African Journal of Economic and Management Sciences (SAJEMS). 2019;**22**(1):8

[16] Sousa AL, Ribeiro T, Relvas S, Barbosa-Póvoa A. Using machine learning for enhancing the understanding of bullwhip effect in the oil and gas industry. Machine Learning and Knowledge Extraction. 2019;**1**:994-1012. DOI: 10.3390/make1030057

[17] Tamim WA, Nawaz RR. Supply Chain Management: Reducing the Bullwhip Effect in SME's. Market Forces. 2017;**12**(1):55

[18] Mafini C, Omoruyi O. Logistics benefits and challenges: The case of SMEs in a South African local municipality. The Southern African Journal of Entrepreneurship and Small Business. 2013;**6**(1):145

[19] Soroka A, Liu Y, Han L, Haleem MS. Big data driven customer insights for SMEs in redistributed manufacturing. Procedia CIRP. 2017;**63**:692-697

[20] Bryman A. Social Research Methods. 4th ed. Oxford: Uk Oxford University Press; 2012

[21] Smart Vision Europe. What Is the CRISP-DM methodology. Smart Vision - Europe. 2017. [Internet] Available from: https://www.sv-europe.com/crisp-dm-methodology/

[22] Rodrigues I. CRISP-DM methodology leader in data mining and big data [Internet]. Medium. 2020. [cited 2021 Jan 1]. Available from: https://towardsdatascience.com/crisp-dm-methodology-leader-in-data-mining-and-big-data-467efd3d3781

[23] Jim F. Regression analysis: An intuitive guide. Statistics By Jim. 2019. [ebook] [Internet] [cited 2020 Jan 1]. Available from: https://statisticsbyjim.selz.com/item/regression-analysis-an-intuitive-guide

[24] Akoglu H. User's guide to correlation coefficients. Turkish Journal of Emergency Medicine. 2018;**18**(3):91-93

[25] James G, Witten D, Hastie T, Tibshirani R. An Introduction to Statistical Learning. New York, Ny: Springer; 2013

[26] Frost J. Multicollinearity in regression analysis: Problems, detection, and solutions - statistics by Jim. Statistics by Jim. 2017. [cited 2021 Jan 1]. [Internet] Available from: https://statisticsbyjim.com/regression/multicollinearity-in-regression-analysis/

[27] Sobolewska E. RPubs - principal component regression [Internet]. rpubs.com. 2019. [cited 2022 Jan 1]. Available from: https://rpubs.com/esobolewska/pcr-step-by-step

[28] Gracey M. When B2B data is personal data and what that means with the GDPR. Medium. 2017. [Internet] [cited 2020 Jan 1]. Available from: https://medium.com/@digital_compliance/when-b2b-data-is-personal-data-and-what-that-means-with-the-gdpr-d4223ea74e09

[29] Muller L. The South African toy market - a country divided, yet incredibly promising. Seekeing Alpha. 2017. [Internet] Available from: https://seekingalpha.com/article/4085341-south-african-toy-market-country-divided-yet-incredibly-promising

[30] Bellégo C, Benatia D, Pape LD. Dealing with logs and zeros in regression models [Internet]. papers.ssrn.com. 2021. [cited 2022 Jan 1]. Available from: https://papers.ssrn.com/sol3/papers.cfm?abstract_id=3444996

[31] Rajput P. Exploratory factor analysis [Internet]. rstudio-pubs-static.s3.ama

zonaws.com. 2018. [cited 2022 Jan 1]. Available from: https://rstudio-pubs-static.s3.amazonaws.com/376139_e9adaefdf4594a79a54a3f87ff4852d6.html#:~:text=Factor%20Analysis%20Model%20Adequacy

[32] Prabhakaran S. 10 Assumptions of Linear Regression - Full List with Examples and Code [Internet]. r-statistics.co. 2016. [cited 2020 Jan 1]. Available from: http://r-statistics.co/Assumptions-of-Linear-Regression.html

[33] Han J, Kamber M. Data Mining: Concepts and Techniques. 3rd ed. Waltham, MA, USA: Elsevier; 2012

[34] Gautam A, Singh V. Parametric versus non-parametric time series forecasting methods: A review. Journal of Engineering Science and Technology Review. 2020;**13**(3): 165-171

[35] Hyndman RJ, Athanasopoulos G. Forecasting: Principles and Practice. 2nd ed OTexts2018. [cited 2021 Jan 1]. Melbourne, Australia: OTexts; 2023. [Internet] Available from: https://otexts.com/fpp2/

[36] DataCamp. Splinefun Function - RDocumentation Stats (version 3.6.2) [Internet]. www.rdocumentation.org. [cited 2022 Jan 1]. Available from: https://www.rdocumentation.org/packages/stats/versions/3.6.2/topics/splinefun

[37] Moufaddal M, Benghabrit A, Bouhaddou I. Big Data Analytics for Supply Chain Management. Cham: Springer; 2018. [cited 2021 Jan 1]. pp. 976-986. Available from:. DOI: 10.1007/978-3-319-74500-8_87

[38] Statistic Solutions. Sample Size Calculation [Internet]. Statisticsolutions.com. [cited 2021 Jan 1]. Available from: https://www.statisticssolutions.com/sample-size-calculation-2/

Chapter 4

Application of Machine Learning in Geotechnical Engineering for Risk Assessment

Ali Akbar Firoozi and Ali Asghar Firoozi

Abstract

Within the domain of geotechnical engineering, risk assessment is pivotal, acting as the linchpin for the safety, durability, and resilience of infrastructure projects. While traditional methodologies are robust, they frequently require extensive manual efforts and can prove laborious. With the onset of the digital era, machine learning (ML) introduces a paradigm shift in geotechnical risk assessment. This chapter delves into the confluence of ML and geotechnical engineering, spotlighting its enhanced predictive capabilities regarding soil behaviors, landslides, and structural resilience. Harnessing modern datasets and rich case studies, we offer an exhaustive examination that highlights the transformative role of ML in reshaping geotechnical risk assessment practices. Throughout our exploration of evolution, challenges, and future horizons, this chapter emphasizes the significance of ML in advancing and transforming geotechnical practices.

Keywords: geotechnical engineering, advanced machine learning applications, comprehensive risk assessment, soil behavior prediction, structural stability, landslide detection, digital revolution in geotechnics, future of risk assessment

1. Introduction

In the vast and evolving landscape of civil engineering, geotechnical engineering holds a pivotal position. For centuries, civilizations have relied on the knowledge and expertise of geotechnical engineers to lay foundations, construct edifices, and shape the built environment. At its core, geotechnical engineering is about understanding the Earth's materials and leveraging that understanding to ensure safety and sustainability in construction endeavors. Yet, with the rapid pace of modernization, urban expansion, and increasing demands on infrastructure, traditional methods of geotechnical assessment have shown signs of strain. The complexities and uncertainties involved in analyzing soil mechanics, earth structures, and foundational behaviors have grown multi-fold. Against this backdrop, the rise of computational technologies and, more recently, the advent of ML, offers a beacon of transformation. Machine learning, characterized by its data-driven approach, pattern recognition, and predictive prowess, intersects with geotechnical engineering's pressing need for more nuanced, efficient, and robust risk assessment tools. The promise lies not just in automating what was traditionally manual but in uncovering insights previously unseen, in predicting failures before they manifest, and in optimizing designs for resilience and longevity. This chapter delves deep into this convergence, exploring the potential, challenges, and future horizons of integrating ML into geotechnical engineering for risk assessment.

1.1 Background

Geotechnical engineering stands as a testament to humankind's quest to master the Earth's materials and harness their properties for infrastructural projects. Tracing its roots back to the dawn of civilization, when the first foundations were laid, the discipline today has expanded beyond its foundational tenets, particularly with the rise of urban centers, intricate transport systems, and monumental architectural marvels [1].

Historically, the emphasis within geotechnical engineering was significantly on empirical and observational methods. Techniques involved comprehensive field investigations, laboratory testing of soil samples, and the use of deterministic models to assess the behavior of earth materials [2]. However, as the urban sprawl began demanding more from the land, the discipline faced increasing challenges, many of which lay beyond the scope of conventional methodologies.

Enter the age of computational advancements. The latter part of the twentieth century witnessed the integration of computational methods into geotechnical engineering, offering more sophisticated ways to analyze and predict earth material behaviors. Yet, with the emergence of the twenty-first century and the computational deluge it brought; it became evident that traditional computational tools were merely a steppingstone to what lay ahead: the union of ML with geotechnical engineering.

Machine learning, a subset of artificial intelligence, has demonstrated remarkable success in various sectors, from finance to healthcare, largely attributed to its prowess in pattern recognition and predictive analysis [3]. For geotechnical engineering, with its complex datasets and myriad variables, the integration of ML can be nothing short of revolutionary.

1.2 Purpose of the study

Navigating this transformative era, our chapter seeks to illuminate the potential of ML in geotechnical engineering, especially within the domain of risk assessment. The union of the computational capabilities of ML with the foundational principles of geotechnical engineering is an avenue yet to be fully explored. We aim to probe this integration, discerning its potential in amplifying risk assessment capabilities, and setting the stage for a new era of predictive geotechnical analysis.

This journey will take us through the very fabric of ML, weaving it with geotechnical datasets, case studies, and real-world applications. From forecasting soil behaviors that traditionally took weeks of lab testing, to real-time monitoring of infrastructural health, and predicting vulnerabilities in massive earth-retaining structures, the applications are as vast as they are groundbreaking.

2. Traditional methods in geotechnical risk assessment

Before the ascent of computational methods and ML, geotechnical engineering primarily relied on traditional methodologies that had been honed over decades of practice. These methodologies, deeply rooted in empirical, observational, and deterministic approaches, served as the bedrock for assessing and mitigating risks associated with earth materials and their interactions with man-made structures. They provided a structured framework, allowing engineers to grapple with the inherently variable and complex nature of the subsurface. Through a blend of field investigations, laboratory tests, and deterministic models, these traditional methods strived to forecast the behavior of soils, rocks, and foundations, shaping the landscape of infrastructure projects around the world. While they have played an instrumental role in the successes of countless projects, the dynamic demands of modern construction and urbanization highlight their constraints and the burgeoning need for more advanced, adaptable tools. This section delves into the essence, intricacies, and challenges of these traditional methodologies, laying the groundwork for understanding the promise that ML brings to the realm of geotechnical risk assessment.

2.1 Overview of traditional risk assessment

Risk assessment in geotechnical engineering is rooted in a combination of observational and empirical methods. Over the years, practitioners have leaned heavily on field investigations, laboratory testing, and the application of deterministic models to understand and predict the behavior of soils, rocks, and other related materials. The crux of these methods lies in assessing how earth materials will respond under different loading and environmental conditions, thereby informing the safety and feasibility of various construction projects.

Historically, these methods have had to balance between being rigorous and pragmatic. Due to the inherent variability of soil and rock properties across different sites, geotechnical engineers have often been tasked with making decisions based on limited data, relying on their expertise and the accumulated knowledge of the field [4].

2.2 Field investigations

The foundation of any geotechnical project is comprehensive field investigation. By gathering firsthand information about the site's subsurface conditions, engineers can make informed decisions about design and construction. Common field tests include:

- *Boring and sampling*: This involves retrieving soil or rock samples from various depths using different boring equipment. These samples are then tested in laboratories to determine their properties.

- *In-situ tests*: Tests like the Standard Penetration Test (SPT) and Cone Penetration Test (CPT) are used to assess soil characteristics directly at the site.

2.3 Laboratory testing

Once samples are collected from the field, they undergo a series of laboratory tests to evaluate their mechanical and physical properties. These tests can include:

- *Shear strength tests*: These tests, like the Direct Shear Test and Triaxial Shear Test, assess the soil's resistance to shearing stresses.

- *Consolidation tests*: Used to determine the compressibility and consolidation properties of soils, aiding in predicting settlement of structures.

2.4 Deterministic models

Geotechnical engineers have traditionally relied on deterministic models to predict the behavior of soils and rocks under specific conditions. These models, rooted in the fundamentals of soil mechanics and rock mechanics, offer mathematical formulations to estimate behaviors such as bearing capacity, slope stability, and soil settlement. A classic example that exemplifies these deterministic models is Terzaghi's Bearing Capacity equation, expressed in Eq. (1). This equation is widely used in foundation design to determine the maximum load a soil can support without failure. The parameters in the equation include the effective cohesion of the soil, unit weight, depth, and effective width of the foundation, along with bearing capacity factors that are intrinsically tied to the soil's internal friction angle.

$$q_u = c'N_c + \gamma D_f N_q + 0.5\gamma B' N_\gamma \tag{1}$$

where: q_u = ultimate bearing capacity of the soil; c' = effective cohesion of the soil; γ = unit weight of the soil; D_f = depth of foundation; B' = effective width of the foundation; N_c, N_q, N_γ = bearing capacity factors, which are functions of the internal friction angle (ϕ') of the soil.

While these traditional methods have been instrumental in advancing the field, they are not devoid of limitations. The next section will delve into some of these constraints, setting the stage for understanding the need for ML in enhancing geotechnical risk assessment.

3. Constraints and limitations of traditional geotechnical risk assessment

The traditional methodologies underpinning geotechnical risk assessment, while historically effective, are not without limitations, especially when considering the intricate, unpredictable nature of soil and rock behavior. Let us delve into these constraints in greater detail.

3.1 Inherent variability of soil and rock

Unlike manufactured materials whose properties can be standardized, soils and rocs present a high degree of variability. Even within a few meters, the geological history, depositional environment, and subsequent processes can significantly alter the mechanical and physical properties of these materials. Traditional methods often use average values or best estimates, which might overlook crucial local variations [5]. This variability means that even with meticulous sampling, unexpected behaviors can emerge, posing challenges in prediction and risk management.

- *Spatial variability*: The spatial distribution of soil and rock properties can vary considerably. Conventional assessment often involves interpolating data between sampling points, but this assumes uniformity between these points, which might not be accurate. **Figure 1** offers a visual insight into the varied layers and inconsistencies in soil composition across a small region. Different colors

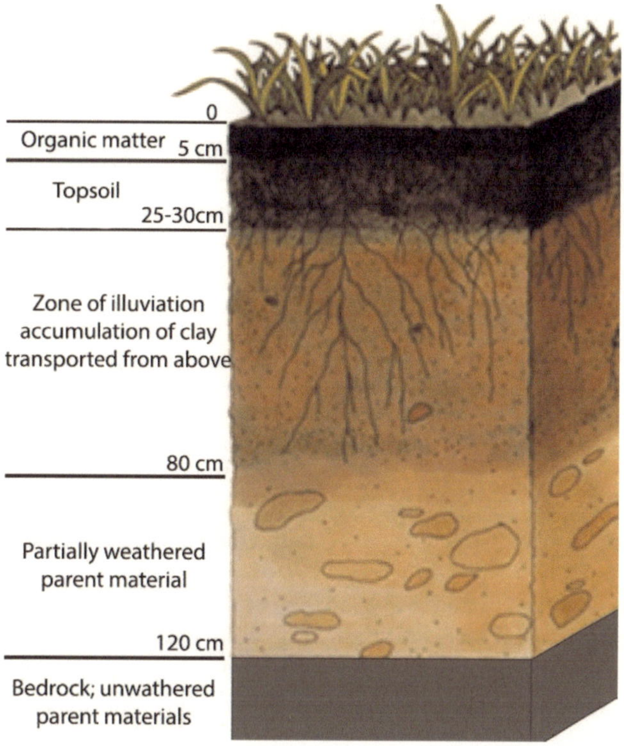

Figure 1.
A graphical representation showcasing the variability of soil layers over a particular region.

represent different soil types, and the uneven layers underline the non-uniform nature of soil stratigraphy.

- *Temporal variability*: Over time, the properties of soils and rocks can change due to factors like weathering, groundwater fluctuations, and biological activities. Traditional methods may not account for these dynamic changes over the lifespan of a structure [6].

3.2 Limitations of deterministic approaches

Geotechnical problems often have numerous variables that interact in intricate ways. Deterministic models, which rely on fixed inputs, can sometimes provide an overly simplistic view. The real-world complexities might not fit neatly into these models, making them less accurate in certain situations [7].

- *Non-linearity in responses*: Many geotechnical problems, such as soil consolidation or slope stability, show non-linear behaviors. Simplifying these into linear models might not capture the true response accurately. Eq. (2) represents the non-linear behavior of soils, especially when subjected to increasing loads. This equation could take the form of a stress-strain curve, commonly used in geotechnical engineering to describe soil behavior under applied stresses.

$$\sigma = E \times \varepsilon^n \tag{2}$$

σ: is the applied stress; E: is the modulus of elasticity, representing the soil's inherent resistance to deformation; ε: is the strain (deformation per unit length); n: is a factor that determines the non-linearity of the stress-strain behavior. For a perfectly elastic material, $n = 1$.

- *Uncertainty handling*: Traditional deterministic methods lack efficient mechanisms to handle and quantify uncertainties. Ignoring these uncertainties might lead to either overly conservative designs or underestimated risks [8].

3.3 Time-consuming laboratory and field tests

While field investigations and lab tests are indispensable, they are often lengthy and resource-intensive. The prolonged duration for results can sometimes hamper the pace of construction projects, especially in environments where rapid decision-making is crucial [9].

- *Cost implications*: Multiple tests, especially when considering depth variability or large sites, can be financially taxing. The extensive equipment, manpower, and subsequent analysis further add to the expenses. **Table 1** provides an overview of standard geotechnical tests and their respective costs. This table could list tests like the Standard Penetration Test (SPT), Triaxial Compression Test, and Direct Shear Test, among others, with associated costs and time durations.

- *Scalability challenges*: For large-scale projects, conducting exhaustive field tests across the entire site might be impractical. Hence, a balance must be struck between coverage and practicality, often leading to potential data gaps.

3.4 Dependence on expert judgment

Much of traditional geotechnical assessment leans heavily on the judgment of experienced engineers. While this expertise is invaluable, it introduces an element of subjectivity, with different experts possibly interpreting data in varied ways [10].

- *Variability in recommendations*: Different experts might arrive at different conclusions given the same data set, leading to variability in design recommendations and potential risks.

- *Over-reliance on past experiences*: While past experiences provide a rich knowledge base, over-reliance on them might deter the exploration of novel, potentially more efficient solutions.

Name	Purpose	Average duration	Approximate cost
Standard Penetration Test (SPT)	Determine soil strength	3 hours	$200
Triaxial Compression Test	Assess soil deformation	6 hours	$400
Direct Shear Test	Measure shear strength	4 hours	$300

Table 1.
Overview of standard geotechnical tests.

3.5 Difficulty in real-time monitoring and prediction

While traditional methods excel in pre-construction assessments, real-time monitoring during and post-construction can be challenging. Continuous monitoring setups, if established, often demand significant resources, and they might not be adept at predicting unforeseen failures swiftly [11].

3.6 Empirical nature of traditional models

A significant portion of traditional geotechnical engineering models is empirically derived. These models, developed from observed behavior in specific conditions, might not universally apply across varied geographies or under different circumstances [12]. While they offer a starting point, relying solely on empirical models might lead to potential inaccuracies.

- *Regional limitations*: Many empirical models were derived from studies in specific regions, reflecting the local geology and environmental conditions. Applying these models to regions with different geological histories or climates might introduce errors. For instance, a model developed in the temperate climates of Europe might not necessarily apply seamlessly to the tropical terrains of Southeast Asia.

- *Aging of empirical data*: As our understanding of geotechnical behavior advances and as technological tools become more sophisticated, older empirical models might become outdated. These models, while still valuable, might not capture the nuances that newer research and technology have unveiled.

3.7 Challenges in large-scale integration

Geotechnical risk assessment often needs to be integrated with other domains, such as structural engineering, hydrology, and environmental science. Traditional methods, often siloed, can sometimes face challenges in this multi-disciplinary integration [13]. Addressing a problem from a purely geotechnical standpoint might overlook interactions and feedback loops from other domains, leading to potential miscalculations.

- *Data compatibility issues:* When interfacing with other domains, data compatibility becomes a challenge. Different fields might use varying metrics, scales, or data formats. Manually harmonizing this data is not only time-consuming but also prone to errors.

- *Complexity in multi-disciplinary communication*: Effective risk assessment in large projects requires seamless communication between different teams. Traditional methods, with their unique terminologies and approaches, might pose barriers in multi-disciplinary communication, leading to potential misunderstandings or oversights.

3.8 Environmental and ethical considerations

With increasing emphasis on sustainable and ethical engineering practices, traditional geotechnical methods face scrutiny. Some methods might involve intrusive site investigations, which can disturb local ecosystems or even local communities. There's an increasing need for methods that are not only technically sound but also environmentally friendly and socially responsible [14].

- *Sustainability concerns:* Traditional risk assessment might not always factor in long-term environmental implications. For example, certain foundation techniques might alter local groundwater flow, leading to unintended environmental consequences in the future.

- *Ethical implications:* Intrusive site investigations or large-scale excavations might disrupt local communities, either by displacing them or by affecting their local environment. Traditional methods need to evolve to ensure that geotechnical work respects both the physical and socio-cultural landscapes.

4. Machine learning: a paradigm shift in geotechnical risk assessment

In the last two decades, the realms of data analytics and computational power have grown exponentially, transcending across a myriad of disciplines. One such beneficiary is the field of geotechnical engineering, which has always grappled with the uncertainties inherent to its subject matter: the earths subsurface. The incorporation of ML techniques is seen as not just an enhancement, but a potential game-changer in deciphering the cryptic terrains and soils below our feet [15].

Table 2 offers a side-by-side comparison of traditional geotechnical risk assessment methods with their ML counterparts. By analyzing their primary features, benefits, and limitations, the table provides a holistic view of how the two approaches fare in various facets of risk evaluation. Notably, ML techniques often present advantages in data processing speeds and predictive accuracy. However, they require vast datasets for optimal functionality. In contrast, traditional methods, while more time-intensive, are backed by tried-and-tested theories and methodologies.

ML technique	Application in geotechnical engineering	Advantages	Limitations
Supervised learning	Soil classification, foundation prediction	Direct mapping of input-output relationships	Requires labeled data
Unsupervised learning	Anomaly detection, soil clustering	Data exploration without predefined labels	Might miss human-defined patterns
Reinforcement learning	Real-time site adjustments, equipment optimization	Dynamic decision making in uncertain environments	Needs simulation or trial environment
deep learning (neural nets)	Complex soil behavior modeling, image recognition	Can model intricate patterns and relationships	Requires large datasets, can be opaque
Transfer learning	Quick model adaptation for new sites	Uses knowledge from previous models/tasks	Might not always transfer effectively
Federated learning	Distributed data training while maintaining privacy	Data privacy and localized training	Might be slower than centralized training

Table 2.
Overview of ML techniques in geotechnical engineering.

4.1 An overview of ML in geotechnical context

Machine Learning, an integral branch of artificial intelligence, thrives on the premise of using data to teach machines how to make decisions, predictions, or classifications without being explicitly programmed for the task. Traditional geotechnical analyses, though rooted in robust scientific principles, often struggled with the vast variability and unpredictability of subsurface conditions. Every construction site, every hillside, every patch of land has its unique geological history and composition. It's in these scenarios that ML excels—by sifting through voluminous datasets, discerning patterns, and predicting geological behavior with a finesse that often surpasses human analysis. These datasets can encompass historical geotechnical reports, real-time monitoring data from sensors, satellite imagery, and even anecdotal evidence from prior construction mishaps or successes [16].

4.2 Advantages over conventional techniques

The very essence of geotechnical engineering revolves around grappling with uncertainties. The Earth, in its eons of existence, has developed intricate layers, fault lines, water tables, and myriad other geological phenomena. Traditional methods, while insightful, often come with constraints tied to their empirical nature and the inherent unpredictability of the subsurface. This is where ML, with its data-driven approach, can offer a fresh perspective.

- *Precision in predictions*: One of the foremost benefits is the refined accuracy ML models bring to the table. Unlike deterministic models, which are limited by predefined parameters, ML models evolved. As they are exposed to more data, their predictive accuracy regarding soil behaviors, landslide susceptibility, or even seismic activities, improves. This dynamic learning curve is indispensable in scenarios like underground tunneling or skyscraper construction, where risks are high, and margins for error are minimal [17].

- *Efficiency through automation*: Traditional geotechnical risk assessments are often labor-intensive. From collecting soil samples to conducting laboratory tests, the process can be prolonged. ML models, once adequately trained, can automate a plethora of these tasks. For instance, with sensors providing real-time data from a construction site, ML algorithms can instantly analyze the data and flag potential anomalies or risks.

- *Adaptability to new data*: The dynamic nature of ML models ensures that they are not static. As fresh data streams in—be it from a new geological survey, updated satellite imagery, or recent seismic activity—these models can be retrained, ensuring that risk assessments are always based on the most current and relevant data [18].

- *Comprehensive data integration*: The versatility of ML is evident in its ability to process and integrate a plethora of data types. Whether it's the chemical composition of a soil sample, infrared imagery from a satellite, or historical data on past landslides in a region, ML algorithms can factor in all these diverse datasets to produce a holistic risk assessment [19].

4.3 Pioneering machine learning techniques in geotechnics

The adaptability of ML techniques means that multiple algorithms and methods find applicability in geotechnical scenarios. Let us delve into some of the most prominent ones.

- *Neural networks*: At the forefront of pattern recognition, neural networks draw inspiration from human brain structures. In a geotechnical context, they have been instrumental in analyzing intricate soil data, deciphering patterns that might be imperceptible through traditional analysis. For instance, predicting how a particular soil type might respond to dynamic loads, like those from an earthquake, becomes more nuanced with neural networks [20].

- *Decision trees and random forests:* While decision trees simplify complex geotechnical decisions by breaking them down into a tree-like model of choices, random forests—ensembles of multiple decision trees—enhance this process's accuracy. For instance, determining the optimal foundation type for a structure in a flood-prone area becomes a more data-driven decision with these algorithms.

- *Support vector machines (SVM)*: SVMs shine in classification problems. In geotechnics, this could translate to categorizing soils based on their bearing capacities or liquefaction potential. Such classifications can be pivotal in decisions related to foundational depths and types [21].

- *Regression analysis*: This technique is particularly valuable when we need to predict a continuous outcome variable based on one or more predictor variables. For instance, using regression analysis, one might predict the rate of soil settlement over time for a particular structure, given certain soil properties and loading conditions.

4.4 Applications of ML in geotechnical risk assessment

Machine learning's true prowess lies in its adaptability and its capability to distill complex patterns from vast datasets. Given the intricate nature of geotechnical engineering, several applications have emerged over the years, revolutionizing traditional risk assessment methods.

- *Landslide susceptibility mapping*: Landslides can be catastrophic, causing significant property damage and loss of life. Predicting their occurrence, based on various factors like soil composition, rainfall data, slope gradient, and human activities, becomes pivotal. ML algorithms, especially neural networks and decision trees, have been employed to analyze these multifaceted datasets, culminating in more accurate landslide susceptibility maps. These maps assist urban planners, especially in hilly terrains, to make informed decisions about infrastructure development and hazard mitigation [22].

- *Foundation behavior prediction*: The foundation is the cornerstone of any infrastructure. Predicting its behavior, especially in variable soil types, becomes imperative. Regression models and Support Vector Machines have found applications here. By analyzing historical data about foundation settlements, tilts, and failures, ML models can predict potential foundation behaviors in given geological conditions. This predictive capability is invaluable in both urban skyscrapers and remote infrastructures like wind turbine foundations [23].

- *Soil classification and characterization*: The classification of soil has always been central to geotechnical studies. Traditional methods, while effective, can be time-consuming. ML, particularly clustering algorithms, has transformed this process. By analyzing various soil properties like grain size, plasticity, and moisture content, ML algorithms can classify soils into various categories, aiding in better design and risk assessment [24].

- *Seismic activity and earthquake prediction*: While the exact prediction of earthquakes remains elusive, significant strides have been made in understanding seismic patterns using ML. Deep learning, a subset of ML, has been instrumental in analyzing seismographs, detecting minor tremors (often imperceptible to human senses), and mapping potential seismic zones. These insights are crucial, especially in earthquake-prone regions, guiding infrastructure development and disaster preparedness [25].

4.5 Challenges in integrating machine learning in geotechnical risk assessment

While the potential of ML in geotechnics is undeniable, it's essential to recognize the inherent challenges in merging these two domains.

- *Data quality and quantity*: ML thrives on data. The accuracy and relevance of ML predictions are directly contingent upon the quality and volume of the data fed to the algorithms. In geotechnical scenarios, acquiring vast datasets that are also accurate can be challenging. Field data is often sparse, and laboratory tests can be inconsistent. Ensuring data reliability becomes paramount [26].

- *Interpretability of models*: ML models, especially complex ones like neural networks, can sometimes act as 'black boxes.' While they might provide accurate predictions, understanding the rationale behind these predictions can be challenging. In critical applications like infrastructure development, stakeholders often require transparent decision-making processes [27].

- *Over-reliance and overfitting*: An over-reliance on ML models without considering the intrinsic uncertainties of geotechnical processes can lead to skewed risk assessments. Similarly, overfitting—a scenario where the ML model is too tailored to the training data—can result in models that perform poorly in real-world scenarios [28].

4.6 Integration of machine learning into current geotechnical practices

Modern geotechnical practices have greatly benefited from the integration of computational tools and methodologies. ML, with its immense capabilities, serves as a natural fit for addressing many complex problems inherent in geotechnical engineering.

- *Pre-processing and data cleansing*: Before any ML model can be trained, the data needs to be prepared, cleansed, and possibly augmented. For geotechnical data, this might involve normalization (scaling all features to a similar range), handling missing values, and even potentially combining multiple datasets. Many geotechnical firms now employ data scientists dedicated to this role, underscoring its significance [29].

- *Automated data collection and integration*: With advancements in sensor technology and IoT (Internet of Things), it's now feasible to collect real-time data from construction sites, drilling rigs, and even deep underground. ML algorithms can integrate this data, offering immediate insights and potentially identifying risks or anomalies in real-time. This proactive approach significantly reduces reaction times in case of unforeseen issues [30].

- *Decision support systems*: For geotechnical engineers, making informed decisions is paramount. By integrating ML models into decision support systems, engineers can simulate various scenarios, forecast potential problems, and make decisions backed by data-driven insights. These systems not only aid in the design phase but also during construction and post-construction monitoring [31].

- *Real-time monitoring and predictive maintenance*: Post-construction, many structures (bridges, tunnels, dams) require consistent monitoring. ML algorithms can analyze the myriad of data points from sensors, detect minute shifts or changes, and predict potential failure points. This shift from reactive to predictive maintenance can save both resources and lives. For instance, if a dam's integrity is at risk, early prediction can lead to timely evacuations and necessary repairs, mitigating potential disasters [32].

4.7 Future directions in geotechnical risk assessment with machine learning

As with any burgeoning technology, the horizon for ML in geotechnical engineering is vast and largely unexplored. The coming years will undoubtedly witness transformative innovations and methodologies.

- *Federated learning for data privacy*: Given the sensitive nature of many infrastructural projects, data privacy is paramount. Federated learning, a form of ML where the model is trained across multiple devices or servers without data centralization, can be a game-changer. This ensures that data never leaves its original location, thus maintaining confidentiality [33].

- *Quantum computing and advanced simulations*: Quantum computing promises unparalleled computational power. In geotechnical engineering, this can lead to simulations of unprecedented accuracy. Combined with ML models, we might soon be looking at almost perfect predictions, especially in complex scenarios like earthquake simulations or underwater tunneling [34].

- *Integration with augmented reality (AR) and virtual reality (VR)*: For on-site engineers and decision-makers, visual data often supersedes numerical data. Integrating ML predictions with AR or VR can provide real-time visual insights.

For instance, using AR glasses, an engineer might see potential soil shifts or weak foundation points overlaid on the actual construction site, aiding immediate decision-making [35].

5. Case studies: machine learning in action

A detailed exploration of specific projects can provide invaluable insights into the practical implications and benefits of integrating ML into geotechnical engineering. This section will delve into real-world applications, emphasizing both successes and challenges faced in the integration process.

5.1 Landslide prediction in the Himalayan region

The Himalayan region is known for its challenging terrains and frequent landslides, particularly during the monsoon season. In a recent project, geotechnical engineers collaborated with data scientists to develop a ML model that would predict potential landslide zones based on various factors such as rainfall, soil moisture, vegetation cover, and slope gradient.

The data was sourced from various remote sensing instruments and ground observations. A combination of supervised and unsupervised learning was employed. The model was trained on past landslide events, with features being the various environmental and geotechnical factors. The outcome was a risk score indicating the likelihood of a landslide occurring in a particular area.

The success of the model was evident when it managed to predict several high-risk zones that were previously not identified using traditional methods. Moreover, the model's real-time data processing capability allowed authorities to take timely evacuation measures, saving numerous lives [36].

5.2 Foundation analysis in urban settings

Urban construction often poses unique challenges, especially when considering the foundation. Given the variable nature of soil and underground utilities in such settings, a ML model was developed to predict the best foundation type (shallow, deep, or pile foundation) for various sites across New York City.

Using a dataset comprising soil samples, underground utility maps, and previous construction projects, the model was trained using supervised learning. The model's recommendations often aligned with geotechnical engineers' judgments, but more importantly, it could identify sites where traditional evaluations were potentially erroneous, thus preventing costly construction errors and delays [37].

5.3 Earthquake damage prediction in Japan

Japan, given its position on the Pacific "Ring of Fire," faces consistent earthquake threats. Accurate prediction of infrastructural damage during earthquakes can save both lives and resources. A project initiated by the University of Tokyo focused on leveraging ML for this very purpose.

They used a dataset encompassing decades of seismic activity, construction details, and post-earthquake damages. Deep learning networks were trained to analyze patterns and predict which structures would likely suffer severe damage during future earthquakes. The model could effectively forecast the probable structural damages during simulations of past major earthquakes, providing valuable insights for urban planning and disaster management. With real-time data from seismic sensors, the model also suggests evacuation measures in vulnerable zones, further enhancing its practicality [38].

5.4 Soil liquefaction analysis in New Zealand

Post the 2011 Christchurch earthquake, there was a dire need to understand and predict soil liquefaction better—a phenomenon where soil loses its strength and stiffness due to an applied stress such as an earthquake, causing it to behave like a liquid. To tackle this, geotechnical engineers teamed up with data scientists in a project funded by the New Zealand government.

The team gathered extensive data on soil types, moisture content, and historical earthquake impacts across various regions in New Zealand. Utilizing a combination of supervised and unsupervised learning, they developed a model that could predict regions susceptible to liquefaction. The results were groundbreaking, enabling city planners to devise strategies to mitigate potential damages and protect key infrastructures from future seismic events [39].

5.5 Tunnel construction monitoring in the Swiss Alps

Tunnel construction in mountainous regions is an arduous task, with a plethora of challenges ranging from unpredictable soil behavior to the risk of water ingress. During the construction of a new railway tunnel in the Swiss Alps, ML models were employed to optimize the process.

Data from sensors embedded in the drilling machines and the tunnel walls, combined with geological surveys, fed into an ML model. This model continuously analyzed the data, predicting areas of potential water ingress or unstable soil layers. The predictions allowed engineers to adjust their drilling strategy in real-time, preventing potential cave-ins and ensuring the safety of the workers [40].

5.6 Detection of sinkholes in Florida

Florida is renowned for its limestone terrain, which is susceptible to the formation of sinkholes. These phenomena pose significant risks to infrastructure and residents. The Florida Geological Survey and the University of Florida collaborated on a project to harness ML in the early detection of sinkholes.

They amassed data involving underground water levels, seismic activity, and prior sinkhole occurrences. Using supervised learning, they built a model to predict potential sinkhole formations based on anomalies in the data. With an accuracy rate of over 90%, this tool became instrumental for urban planners and property developers in avoiding areas at risk and planning remedial measures for existing structures [41].

5.7 Slope stability in the Andean region

The Andean region, with its steep terrains and frequent rainfall, is prone to landslides and slope failures. The local government, in conjunction with geotechnical consultants, integrated ML to assess and predict slope stability.

A neural network model was trained using data on rainfall patterns, soil types, slope gradients, and vegetation cover. By continually assessing these parameters, the model offered real-time evaluations of slope stability, suggesting when and where interventions might be needed. This proactive approach has drastically reduced landslide incidences in critical infrastructure zones [42].

5.8 Groundwater contamination prediction in industrial regions

Groundwater contamination in industrial zones is a growing concern worldwide. In a pioneering effort in Germany, researchers developed a ML model to predict areas at risk of contamination based on industrial activities, soil permeability, and underground water flows.

This model utilized a combination of unsupervised learning for anomaly detection and supervised learning for predictive analytics. It highlighted zones at high risk and recommended changes in industrial activities or enhanced containment measures. This predictive tool has since become a standard reference for environmental clearance of new industrial projects in the region [43].

5.9 Reinforcement learning in automated drilling

Automated drilling systems have gained prominence in large-scale geotechnical projects. An ongoing research at Stanford University focuses on integrating reinforcement learning into automated drilling systems. The objective is to allow the system to learn from its environment in real-time and make decisions that optimize drilling efficiency while ensuring safety. Initial results indicate a potential reduction in project timeframes by up to 15% and a significant decrease in equipment wear and tear [44].

6. Challenges and future directions in integrating ML into geotechnical risk assessment

The merger of ML with geotechnical risk assessment is akin to the confluence of two powerful rivers; while the combined force can carve new paths and offer unparalleled advantages, it also brings forth a set of challenges that are unique to their union. With the promises of enhanced predictive power and efficient analysis, ML methods beckon a future of transformative geotechnical practices. However, the path is not devoid of obstacles. Navigating issues related to data quality, model transparency, scalability, and practical implementation demands collaborative efforts from both ML practitioners and geotechnical engineers. This section delves deep into these challenges, attempting not just to highlight them but also to offer a perspective on potential solutions and the road ahead.

6.1 The challenge of data collection and pre-processing

In the realm of ML, the axiom "Garbage in, garbage out" stands unequivocally true. For any ML model to be effective, especially in the meticulous domain of geotechnical engineering, the data fed into the system needs to be both relevant and precise. Historically, geotechnical data has been scattered, inconsistent, and sometimes incomplete. The reasons span from diverse measurement techniques to regional variations in data recording and even economic constraints that limit extensive data collection.

Collecting robust, comprehensive, and standardized data is a monumental task, especially for regions where geotechnical studies have been historically underfunded or overlooked. Moreover, once data is collected, the pre-processing stage can be equally daunting. Raw data often comes with noise, outliers, or missing values. Cleaning this data, normalizing it, and making it suitable for ML models demands significant effort and expertise. The transformation of raw geotechnical data into a format that is machine-readable and conducive to accurate predictions remains a significant hurdle [45].

6.2 Model interpretability and trust

Another formidable challenge is the black-box nature of many advanced ML models. Geotechnical engineers, by the very nature of their work, are inclined to trust models and systems that provide a clear cause-and-effect relationship. When a ML model, such as a deep neural network, produces a prediction or a risk assessment, the path to that conclusion is not always transparent. The opacity of these models can lead to hesitation in their adoption, especially in high-stake scenarios were understanding the 'why' behind a prediction can be as crucial as the prediction itself.

This challenge is not insurmountable. Recent advances in the field of explainable AI (XAI) aim to make ML models more interpretable. By offering insights into the decision-making process of the model, these tools are striving to bridge the trust gap. However, integrating XAI into geotechnical risk assessments is still a work in progress, and widespread trust in ML outcomes remains a goal for the future [46].

6.3 Scalability and real-time processing

While ML models excel in handling vast datasets and intricate computations, scalability in real-time environments remains a challenge. Geotechnical risk assessments often demand instantaneous decisions, especially in scenarios like live monitoring of landslides or the structural integrity of infrastructures in earthquake-prone areas. The larger and more complex the ML model, the greater computational power it requires, which can sometimes be a bottleneck in delivering real-time insights.

Furthermore, with the ongoing collection of data, models need to be periodically retrained or fine-tuned. Ensuring this happens seamlessly without disrupting real-time assessments is a challenge that engineers and data scientists grapple with [47].

6.4 Integration with existing systems

Most geotechnical firms and institutions have existing systems in place for risk assessment. These systems, built over years or even decades, are deeply embedded into their operational workflows. The integration of ML models into these legacy systems is no trivial task. It demands not just technical adaptations but also a cultural shift. Training personnel, adapting to new decision-making paradigms, and ensuring that the integration does not disrupt ongoing operations are all significant challenges [48].

6.5 Ethical considerations and accountability

With the advent of ML in risk assessment, ethical dilemmas surface. Who bears the responsibility if an ML model's prediction goes awry leading to infrastructural damage

Challenge	Implication	Potential solutions
Data collection and pre-processing	Inconsistent and incomplete datasets	Standardization, enhanced funding, and advanced sensors
Model interpretability and trust	Reluctance to adopt opaque models	Integration of Explainable AI (XAI) tools
Scalability and real-time processing	Delays in decision-making in critical situations	Optimized algorithms and distributed computing
Integration with existing systems	Disruptions in current workflows	Training programs and phased integration
Ethical considerations and accountability	Dilemmas over responsibility in case of model failures	Clear legal frameworks and guidelines

Table 3.
Summary of challenges in integrating ML into geotechnical risk assessment.

or, worse, loss of life? The automation of decisions, especially in critical areas like geotechnical risk, brings forth questions of accountability. Establishing clear guidelines, standards, and legal frameworks for the deployment and outcomes of ML models in geotechnical engineering is an imperative challenge that professionals and policymakers need to address collectively [49].

Table 3 succinctly encapsulates the primary challenges encountered when incorporating ML into geotechnical risk assessment. By outlining the implications of each challenge, it offers a clear view of the hurdle's professionals face in this interdisciplinary endeavor. Moreover, the 'Potential Solutions' column highlights proactive steps and strategies that can address, if not completely overcome, these challenges. This table is essential as it not only underscores the problems but also emphasizes that solutions, although demanding, are within reach.

7. Potential of advanced ML techniques in geotechnical risk assessment

The amalgamation of ML techniques with geotechnical risk assessment is not just about addressing challenges; it's also a doorway to new possibilities that were previously unattainable. Advanced ML techniques, including deep learning, reinforcement learning, and transfer learning, open up avenues that can revolutionize how geotechnical risks are predicted, analyzed, and mitigated. This section delves into these advanced techniques, exploring their potential applications and the transformative impacts they can bring to the field of geotechnical engineering.

7.1 Deep learning and soil behavior analysis

Deep learning, a subset of ML, employs neural networks with many layers (deep neural networks) to analyze various types of data. In the context of geotechnical engineering, deep learning can be instrumental in understanding complex soil behaviors that have traditionally been difficult to model. For instance, the non-linear behavior of certain soils under varied loading conditions can be efficiently modeled using deep learning techniques, providing insights that are closer to real-world scenarios [50].

7.2 Reinforcement learning for optimal infrastructure placement

Reinforcement learning (RL) is a type of ML where an agent learns by interacting with its environment and receiving feedback in the form of rewards or penalties. When applied to geotechnical risk assessment, RL can be used to determine the optimal placement of infrastructure elements, like pillars or retaining walls, in challenging terrains. The RL agent can simulate numerous placements, learn from the results, and eventually propose a design that minimizes risk while optimizing utility [51].

7.3 Transfer learning and global risk prediction models

Transfer learning is the practice of applying knowledge gained from one task to a different, yet related, task. In geotechnical terms, this means that an ML model trained on data from one geographic region might be adapted to make predictions in another region, given some fine-tuning. This approach can be particularly beneficial in areas where data is scarce, allowing engineers to leverage global datasets for local risk predictions [52].

7.4 Generative adversarial networks (GANs) for simulating soil profiles

GANs, in their unique design, have revolutionized data synthesis. In geotechnical engineering, the challenge has always been the unpredictable nature of soil profiles over vast stretches. Traditional methods might provide limited insights based on point sampling, but GANs open up an avenue where synthetic yet scientifically accurate soil profiles can be generated.

Take the instance of a construction company planning to build a long tunnel. The soil profile, composition, and characteristics might vary drastically over small distances. Instead of extensive and expensive physical samplings, GANs, trained on a diverse range of soil datasets, can simulate potential profiles. These profiles would then guide engineers to anticipate challenges and optimize construction methods accordingly. Zhang et al. [53] study highlighted a 30% reduction in unexpected geotechnical challenges during tunnel constructions using GAN-generated soil profiles.

7.5 Time series forecasting for predicting landslide movements

Historically, predicting the exact moment or scale of landslides was analogous to predicting earthquakes, fraught with uncertainties. However, time series forecasting, especially when applied to data-rich environments, has altered this landscape. By continually monitoring soil movements, moisture levels, and other critical parameters, and then feeding this data into time series models, accurate predictions about potential landslide activities can be made.

In the Himalayan region, known for its treacherous landslides, a study by Al-Najjar et al. [54] implemented time series forecasting models in 10 critical regions. The results were startling. Early warnings were issued in seven regions, allowing authorities ample time to evacuate or secure areas, thus averting potential disasters.

7.6 Ensemble learning for enhanced prediction accuracy

The complexity of geotechnical parameters makes it an ideal candidate for ensemble learning. No single model can predict with absolute certainty, given the myriad variables. But when multiple models, each with its strength, are combined, the prediction accuracy elevates significantly.

Consider the challenge of predicting the stability of a retaining wall. Factors like soil type, moisture, load, previous movements, etc., play a role. While a deep learning model might excel in understanding soil behavior, a reinforcement learning model could provide insights into optimal load adjustments. Ensemble learning brings these models together, offering a comprehensive prediction. Krechowicz and Krechowicz [55] showcased that ensemble models, on average, improved prediction accuracies by 18% over singular models in complex geotechnical scenarios.

Table 4 offers a comparative overview of the three advanced ML techniques discussed in this section, highlighting their primary use cases in geotechnical engineering. Additionally, the table provides insights into the percentage increase in prediction accuracy where applicable and references notable studies associated with each technique. Such a table provides readers with a succinct summary, allowing for quick cross-referencing and comprehension.

In summary advanced ML techniques, from GANs to ensemble learning, offer transformative approaches to age-old geotechnical challenges. These techniques not only enhance predictive accuracy but also provide tools to simulate, analyze, and optimize in ways previously deemed unattainable. As the integration of these methods with geotechnical engineering deepens, we are on the precipice of a new era, one where risks are better understood, anticipated, and mitigated, safeguarding infrastructure and lives alike.

8. Practical integration of ML in geotechnical risk assessment

As the possibilities of ML in geotechnical engineering come to the forefront, it's vital to understand the practical steps for integrating these powerful tools. While the theoretical potentials are promising, actual integration requires a systematic approach to ensure optimal results.

8.1 Data collection and preprocessing in geotechnical engineering

Before applying any ML model, the quality and quantity of data are paramount. In geotechnical engineering, collecting the right data can be a daunting task due to the

Technique	Primary use case	Prediction accuracy increase (%)	Notable study
Generative adversarial networks	Simulating soil profiles	30% (in tunnel constructions)	Zhang et al. [53]
Time series forecasting	Predicting landslide movements	Not quantified	Al-Najjar et al. [54]
Ensemble learning	Comprehensive geotechnical risk predictions	18%	Krechowicz and Krechowicz [55]

Table 4.
Comparative analysis of advanced ML techniques in geotechnical engineering.

inherent variability of natural conditions. Nonetheless, advanced sensors, remote sensing technologies, and geotechnical investigations have enabled the collection of vast datasets. It's crucial to preprocess this data, removing outliers, handling missing values, and ensuring that the dataset is representative of the diverse conditions a project might encounter [56].

- *Advanced sensing technologies*: Recent developments in sensor technologies, including piezometers, inclinometers, and extensometers, have facilitated real-time data collection, capturing minute changes in soil mechanics and groundwater pressures [57].

- *Remote sensing and GIS integration*: The combination of satellite imagery, LIDAR, and Geographic Information Systems (GIS) allows for the large-scale assessment of geotechnical properties across vast terrains. This amalgamation aids in identifying potential risk zones even before detailed on-site investigations begin [58].

- *Data cleaning and preprocessing*: Once collected, data undergoes rigorous preprocessing. This involves normalizing scales, handling missing or inconsistent data, and using techniques like Principal Component Analysis (PCA) to reduce dimensionality, ensuring efficient training of ML models [59].

8.2 Model selection and training

Choosing the right model for the task is essential. While advanced techniques like GANs or ensemble learning offer excellent results in specific scenarios, simpler models might suffice for others. Training the selected model using geotechnical datasets ensures that it becomes attuned to the nuances of the field. Regular model validation and iterative training are essential to maintain its accuracy and relevance [60].

- *Criteria for model selection*: Factors like the nature of the data (continuous, categorical), the objective (classification, regression), and the availability of labeled data (supervised vs. unsupervised learning) dictate model selection [61].

- *Regularization techniques*: Overfitting is a significant concern in geotechnical applications due to the natural variability in data. Techniques such as Ridge, Lasso, and Elastic Net regularization are employed to counteract this, making models more generalizable [62].

- *Model validation*: Techniques such as k-fold cross-validation are used to assess model performance on different subsets of data, ensuring its robustness [63].

8.3 Post-model analysis and interpretation

After training, it's not just about obtaining results; it's about interpreting them. ML models, especially the more complex ones, can sometimes act as "black boxes." However, tools like SHAP (SHapley Additive exPlanations) values or LIME (Local Interpretable Model-agnostic Explanations) can help decipher these model outputs.

They enable geotechnical engineers to understand the decision-making process of the algorithm, ensuring that the insights are not just accurate but also actionable [64].

- *Importance of explainable AI (XAI)*: While traditional ML models have been criticized for their opacity, the emerging field of XAI seeks to bridge this gap. Techniques within XAI aim to provide clarity on how decisions are made within an algorithm, ensuring that professionals can trust and act on these insights [65].

- *Real-time monitoring and updates*: Post-deployment, it's imperative that models continue to learn from new data. With the advent of IoT devices in geotechnical sites, continuous feedback loops can be established, allowing for real-time model updates [66].

In summary, the integration of ML in geotechnical risk assessment is a multidimensional task, spanning from meticulous data collection to real-time model adaptation. The roadmap, while complex, promises an evolution in risk assessment, ushering in an era of increased safety, efficiency, and innovation in geotechnical engineering.

9. Conclusions and future prospects

The interfusion of ML techniques with geotechnical engineering has charted a new trajectory in the domain of risk assessment. The myriad of applications, ranging from predicting soil failures and landslides to assessing structural stability, underscores the immense potential of this integration.

- *Revisiting traditional methods*: Traditional geotechnical risk assessment approaches, while robust and well-tested, are being redefined in the light of ML. These traditional methods often were labor-intensive, time-consuming, and occasionally fell short in terms of predictive accuracy. ML, with its ability to process vast datasets, offers a more dynamic, precise, and expedited risk assessment, making it a formidable tool in the geotechnical realm.

- *Challenges in integration*: Despite the promises, integrating ML into geotechnical engineering is not without challenges. Issues related to data privacy, the accuracy of predictions in varying geological conditions, and the need for continuous model training highlight some of the existing limitations. It is imperative for researchers and professionals to address these challenges head-on, ensuring that ML-driven solutions remain effective and reliable.

- *Future directions*: The future seems radiant for the convergence of ML and geotechnical engineering. With the emergence of more advanced algorithms and increased computational capacities, the applications are only expected to expand. We foresee a shift towards real-time monitoring and predictions, wherein sensors placed at strategic locations would relay information instantaneously to ML models, offering almost immediate risk assessments.

- *Emphasis on collaboration*: One of the key takeaways from our exploration is the pressing need for collaboration. Data scientists, ML experts, geotechnical engineers, and urban planners must join forces, pooling their expertise to harness the full potential of ML in geotechnical risk assessment.

- *Evolving educational curricula*: As the dynamics of the industry change, so should the educational paradigms. There's a rising demand for professionals who are adept in both geotechnical principles and ML algorithms. Universities and institutions must revisit their curricula, ensuring they produce professionals ready for this interdisciplinary challenge.

In essence, the realm of geotechnical risk assessment is on the cusp of a transformative phase, powered by the dynamism of ML. While challenges exist, the collaborative efforts of professionals across domains and the incessant advancement in technology promise a future where geotechnical risk assessments are more accurate, swift, and actionable.

Acknowledgements

The authors would like to extend their heartfelt gratitude to the Department of Civil Engineering at University of Botswana for their unwavering support and resources provided during the compilation of this chapter. Furthermore, we are deeply grateful to the countless geotechnical engineers and ML experts whose pioneering work in the field laid the foundation for our research. Their relentless pursuit of knowledge continues to inspire us. Lastly, a special mention to all the anonymous reviewers for their insightful critiques and recommendations, which greatly enhanced the quality of this chapter.

Conflict of interest

The authors declare no potential conflicts of interest concerning the research, authorship, and publication of this chapter. All research and data compilations were conducted with integrity and transparency, ensuring objectivity throughout. All relevant permissions and clearances were sought for the case studies and data sets utilized, ensuring there was no breach of proprietary or confidential information.

Author details

Ali Akbar Firoozi* and Ali Asghar Firoozi
Department of Civil Engineering, Faculty of Engineering and Technology, University of Botswana, Gaborone, Botswana

*Address all correspondence to: a.firoozi@gmail.com

References

[1] Briaud JL. Geotechnical Engineering: Unsaturated and Saturated Soils. Canada: John Wiley & Sons; 2023

[2] Phoon KK, Cao ZJ, Ji J, Leung YF, Najjar S, Shuku T, et al. Geotechnical uncertainty, modeling, and decision making. Soils and Foundations. 2022; **62**(5):101189. DOI: 10.1016/j.sandf.2022.101189

[3] Dong S, Wang P, Abbas K. A survey on deep learning and its applications. Computer Science Review. 2021;**40**: 100379. DOI: 10.1016/j.cosrev.2021.100379

[4] Phoon KK, Ching J, Wang Y. Managing risk in geotechnical engineering–from data to digitalization. In: Proc., 7th Int. Symp. on Geotechnical Safety and Risk (ISGSR 2019). 2019. pp. 13-34

[5] Olaiz AH, Zapata CE, Soltanpour Y. A Bayesian forecast framework for climatic parameters in geotechnical modeling. In: Geo-Risk. 2023. pp. 88-97

[6] Uzielli M, Lacasse S, Nadim F, Phoon KK. Soil variability analysis for geotechnical practice. Characterization and Engineering Properties of Natural Soils. 2006;**3**:1653-1752

[7] Phoon KK, Kulhawy FH. Characterization of geotechnical variability. Canadian Geotechnical Journal. 1999;**36**(4):612-624. DOI: 10.1139/t99-038

[8] Wang L, Wu C, Tang L, Zhang W, Lacasse S, Liu H, et al. Efficient reliability analysis of earth dam slope stability using extreme gradient boosting method. Acta Geotechnica. 2020;**15**: 3135-3150. DOI: 10.1007/s11440-020-00962-4

[9] Atkinson J. An Introduction to the Mechanics of Soils and Foundations: Through Critical State Soil Mechanics. United States: McGraw-Hill Book Company (UK) Ltd.; 1993

[10] Mayne PW, Coop MR, Springman SM, Huang AB, Zornberg JG. Geomaterial behavior and testing. In: Proceedings of the 17th International Conference on Soil Mechanics and Geotechnical Engineering, Egypt. Vol. 1, 2, 3 and 4. IOS Press; 2009. pp. 2777-2872. DOI: 10.3233/978-1-60750-031-5-2777

[11] Brandl H. Energy foundations and other thermo-active ground structures. Géotechnique. 2006;**56**(2):81-122. DOI: 10.1680/geot.2006.56.2.81

[12] Madhusudhan BR, Boominathan A, Banerjee S. Effect of specimen size on the dynamic properties of river sand and rubber tire shreds from cyclic triaxial and cyclic simple shear tests. In geotechnical characterization and modelling. In: Proceedings of IGC. Springer Singapore; 2020, 2018. pp. 453-465. DOI: 10.1007/978-981-15-6086-6_37

[13] Ferrario MF, Bonadeo L, Brunamonte F, Livio F, Martinelli E, Michetti AM, et al. Late quaternary environmental evolution of the Como urban area (northern Italy): A multidisciplinary tool for risk management and urban planning. Engineering Geology. 2015;**193**:384-401. DOI: 10.1016/j.enggeo.2015.05.013

[14] Terzaghi K, Peck RB, Mesri G. Soil Mechanics in Engineering Practice. Canada: John Wiley & Sons; 1996

[15] Zhang W, Li H, Li Y, Liu H, Chen Y, Ding X. Application of deep learning

algorithms in geotechnical engineering: A short critical review. Artificial Intelligence Review. 2021;**54**(8):5633-5673. DOI: 10.1007/s10462-021-09967-1

[16] Abbaszadeh Shahri A, Kheiri A, Hamzeh A. Subsurface topographic modeling using geospatial and data driven algorithm. ISPRS International Journal of Geo-Information. 2021;**10**(5): 341. DOI: 10.3390/ijgi10050341

[17] Tabarsa A, Latifi N, Osouli A, Bagheri Y. Unconfined compressive strength prediction of soils stabilized using artificial neural networks and support vector machines. Frontiers of Structural and Civil Engineering. 2021; **15**:520-536. DOI: 10.1007/s11709-021-0689-9

[18] Phoon KK, Zhang W. Future of machine learning in geotechnics. Georisk: Assessment and Management of Risk for Engineered Systems and Geohazards. 2023;**17**(1):7-22. DOI: 10.1080/17499518.2022.2087884

[19] Dodangeh E, Choubin B, Eigdir AN, Nabipour N, Panahi M, Shamshirband S, et al. Integrated machine learning methods with resampling algorithms for flood susceptibility prediction. Science of the Total Environment. 2020;**705**: 135983. DOI: 10.1016/j.scitotenv.2019.135983

[20] Deng X, Liu P, Liu X, Wang R, Zhang Y, He J, et al. Geospatial big data: New paradigm of remote sensing applications. IEEE Journal of Selected Topics in Applied Earth Observations and Remote Sensing. 2019;**12**(10): 3841-3851. DOI: 10.1109/JSTARS.2019.2944952

[21] Zhou J, Huang S, Wang M, Qiu Y. Performance evaluation of hybrid GA–SVM and GWO–SVM models to predict earthquake-induced liquefaction potential of soil: A multi-dataset investigation. Engineering with Computers. 2022;**38**(5):4197-4215. DOI: 10.1007/s00366-021-01418-3

[22] Merghadi A, Yunus AP, Dou J, Whiteley J, ThaiPham B, Bui DT, et al. Machine learning methods for landslide susceptibility studies: A comparative overview of algorithm performance. Earth-Science Reviews. 2020;**207**: 103225. DOI: 10.1016/j.earscirev.2020.103225

[23] Goodell JW, Kumar S, Lim WM, Pattnaik D. Artificial intelligence and machine learning in finance: Identifying foundations, themes, and research clusters from bibliometric analysis. Journal of Behavioral and Experimental Finance. 2021;**32**:100577. DOI: 10.1016/j.jbef.2021.100577

[24] Vij A, Vijendra S, Jain A, Bajaj S, Bassi A, Sharma A. IoT and machine learning approaches for automation of farm irrigation system. Procedia Computer Science. 2020;**167**:1250-1257. DOI: 10.1016/j.procs.2020.03.440

[25] Szakács A. Precursor-based earthquake prediction research: Proposal for a paradigm-shifting strategy. Frontiers in Earth Science. 2021;**8**: 548398. DOI: 10.3389/feart.2020.548398

[26] Chen J, Vissinga M, Shen Y, Hu S, Beal E, Newlin J. Machine learning–based digital integration of geotechnical and ultrahigh–frequency geophysical data for offshore site characterizations. Journal of Geotechnical and Geoenvironmental Engineering. 2021; **147**(12):04021160. DOI: 10.1061/(ASCE)GT.1943-5606.0002702

[27] Bhattacharya S, Demirci HE, Nikitas G, Prakhya GKV, Lombardi D, Alexander NA, et al. Chapter 11 - Physical modeling of interaction

problems in geotechnical engineering. Modeling in Geotechnical Engineering. Academic Press; 2021. pp. 205-256. DOI: 10.1016/B978-0-12-821205-9.00017-4. ISBN 9780128212059

[28] Pan Q, Qu X, Liu L, Dias D. A sequential sparse polynomial chaos expansion using Bayesian regression for geotechnical reliability estimations. International Journal for Numerical and Analytical Methods in Geomechanics. 2020;**44**(6):874-889. DOI: 10.1002/nag.3044

[29] Kim HJ, Mawuntu KBA, Park TW, Kim HS, Park JY, Jeong YS. Spatial autocorrelation incorporated machine learning model for geotechnical subsurface modeling. Applied Sciences. 2023;**13**(7):4497. DOI: 10.3390/app13074497

[30] Carri A. Innovative application of iot technologies to improve geotechnical monitoring tools and early warning performances. In: Critical Thinking in the Sustainable Rehabilitation and Risk Management of the Built Environment: CRIT-RE-BUILT. Proceedings of the International Conference; November 7-9, 2019, Iași, Romania. Switzerland: Springer Nature; 2020. p. 142

[31] Mahdi IM, Ebid AM, Khallaf R. Decision support system for optimum soft clay improvement technique for highway construction projects. Ain Shams Engineering Journal. 2020;**11**(1): 213-223. DOI: 10.1016/j.asej.2019.08.007

[32] Hallaji SM, Fang Y, Winfrey BK. Predictive maintenance of pumps in civil infrastructure: State-of-the-art, challenges and future directions. Automation in Construction. 2022;**134**: 104049. DOI: 10.1016/j.autcon.2021.104049

[33] Yang Y, Lu Y, Mei G. A federated learning based approach for predicting landslide displacement considering data security. Future Generation Computer Systems. 2023;**149**:184-199. DOI: 10.1016/j.future.2023.07.021

[34] Criekemans D. Chapter 2 'Geotechnical Ensembles': How new technologies change geopolitical factors and contexts in economy, energy and security. Geopolitics and International Relations. Leiden, The Netherlands: Brill | Nijhoff; 2021. DOI: 10.1163/9789004432086_004

[35] Rodríguez Piedrabuena A. Feasibility study of using augmented reality in geotechnical site inspection [Bachelor's Thesis]. Spain: Universitat Politècnica de Catalunya; 2021. Available from: http://hdl.handle.net/2117/358184

[36] Riaz MT, Basharat M, Brunetti MT. Assessing the effectiveness of alternative landslide partitioning in machine learning methods for landslide prediction in the complex Himalayan terrain. Progress in Physical Geography: Earth and Environment. 2023;**47**(3):315-347. DOI: 10.1177/03091333221113660

[37] Seyedzadeh S, Rahimian FP, Oliver S, Rodriguez S, Glesk I. Machine learning modelling for predicting non-domestic buildings energy performance: A model to support deep energy retrofit decision-making. Applied Energy. 2020; **279**:115908. DOI: 10.1016/j.apenergy.2020.115908

[38] Velasco Herrera VM, Rossello EA, Orgeira MJ, Arioni L, Soon W, Velasco G, et al. Long-term forecasting of strong earthquakes in North America, South America, Japan, southern China and northern India with machine learning. Frontiers in Earth Science. 2022;**10**:905792. DOI: 10.3389/feart.2022.905792

[39] Quigley MC, Saunders W, Massey C, Van Dissen R, Villamor P, Jack H, et al. The utility of earth science information in post-earthquake land-use decision-making: The 2010–2011 Canterbury earthquake sequence in Aotearoa New Zealand. Natural Hazards and Earth System Sciences Discussions. 2020;**2020**: 1-35. DOI: 10.5194/nhess-20-3361-2020

[40] Song Q, Wu Y, Xin X, Yang L, Yang M, Chen H, et al. Real-time tunnel crack analysis system via deep learning. IEEE Access. 2019;**7**: 64186-64197. DOI: 10.1109/ACCESS.2019.2916330

[41] Kariminejad N, Mondini A, Hosseinalizadeh A et al. Detection of sinkholes and landslides in a semi-arid environment using deep-learning methods. UAV images, and Topographical Derivatives. 15 May 2023. PREPRINT (Version 1). DOI: 10.21203/rs.3.rs-2847897/v1. Available from: Research Square

[42] Bravo-López E, Fernández Del Castillo T, Sellers C, Delgado-García J. Landslide susceptibility mapping of landslides with artificial neural networks: Multi-approach analysis of backpropagation algorithm applying the neuralnet package in Cuenca. Ecuador. Remote Sensing. 2022;**14**(14):3495. DOI: 10.3390/rs14143495

[43] Elzain HE, Chung SY, Venkatramanan S, Selvam S, Ahemd HA, Seo YK, et al. Novel machine learning algorithms to predict the groundwater vulnerability index to nitrate pollution at two levels of modeling. Chemosphere. 2023;**314**: 137671. DOI: 10.1016/j.chemosphere.2022.137671

[44] Sircar A, Yadav K, Rayavarapu K, Bist N, Oza H. Application of machine learning and artificial intelligence in oil and gas industry. Petroleum Research. 2021;**6**(4):379-391. DOI: 10.1016/j.ptlrs.2021.05.009

[45] Zhang W, Gu X, Tang L, Yin Y, Liu D, Zhang Y. Application of machine learning, deep learning and optimization algorithms in geoengineering and geoscience: Comprehensive review and future challenge. Gondwana Research. 2022;**109**:1-17. DOI: 10.1016/j.gr.2022.03.015

[46] Lee J, Azamfar M, Singh J, Siahpour S. Integration of digital twin and deep learning in cyber-physical systems: Towards smart manufacturing. IET Collaborative Intelligent Manufacturing. 2020;**2**(1):34-36. DOI: 10.1049/iet-cim.2020.0009

[47] Lu X, Xu Y, Tian Y, Cetiner B, Taciroglu E. A deep learning approach to rapid regional post-event seismic damage assessment using time-frequency distributions of ground motions. Earthquake Engineering & Structural Dynamics. 2021;**50**(6): 1612-1627. DOI: 10.1002/eqe.3415

[48] Kim HS, Sun CG, Lee MG, Cho HI. Multivariate geotechnical zonation of seismic site effects with clustering-blended model for a city area, South Korea. Engineering Geology. 2021;**294**: 106365. DOI: 10.1016/j.enggeo.2021.106365

[49] Künzler M, Huggel C, Ramírez JM. A risk analysis for floods and lahars: Case study in the Cordillera Central of Colombia. Natural Hazards. Oct 2012;**64**: 767-796

[50] Yaseen ZM. An insight into machine learning models era in simulating soil, water bodies and adsorption heavy metals: Review, challenges and solutions. Chemosphere. 2021;**277**:130126. DOI: 10.1016/j.chemosphere.2021.130126

[51] Zhang P, Yin ZY, Jin YF. Machine learning-based modelling of soil properties for geotechnical design: Review, tool development and comparison. Archives of Computational Methods in Engineering. 2022;**29**(2): 1229-1245. DOI: 10.1007/s11831-021-09615-5

[52] Jena R, Shanableh A, Al-Ruzouq R, Pradhan B, Gibril MBA, Ghorbanzadeh O, et al. An integration of deep learning and transfer learning for earthquake-risk assessment in the Eurasian region. Remote Sensing. 2023; **15**(15):3759. DOI: 10.3390/rs15153759

[53] Zhang C, Liang M, Song X, Liu L, Wang H, Li W, et al. Generative adversarial network for geological prediction based on TBM operational data. Mechanical Systems and Signal Processing. 2022;**162**:108035. DOI: 10.1016/j.ymssp.2021.108035

[54] Fang H, Shao Y, Xie C, Tian B, Shen C, Zhu Y, et al. A new approach to spatial landslide susceptibility prediction in karst mining areas based on explainable Artificial Intelligence. Sustainability. 2023;**8, 15**(4):3094. DOI: 10.3390/su15043094

[55] Krechowicz M, Krechowicz A. Risk assessment in energy infrastructure installations by horizontal directional drilling using machine learning. Energies. 2021;**14**(2):289. DOI: 10.3390/en14020289

[56] Mitelman A, Yang B, Urlainis A, Elmo D. Coupling geotechnical numerical analysis with machine learning for observational method projects. Geosciences. 2023;**13**(7):196. DOI: 10.3390/geosciences13070196

[57] Barzegar M, Blanks S, Sainsbury BA, Timms W. MEMS technology and applications in geotechnical monitoring: A review. Measurement Science and Technology. 2022;**33**(5):052001. DOI: 10.1088/1361-6501/ac4f00

[58] Jena R, Pradhan B, Beydoun G, Al-Amri A, Sofyan H. Seismic hazard and risk assessment: A review of state-of-the-art traditional and GIS models. Arabian Journal of Geosciences. 2020;**13**: 1-21. DOI: 10.1007/s12517-019-5012-x

[59] Kim HS, Ji Y. Three-dimensional geotechnical-layer mapping in Seoul using borehole database and deep neural network-based model. Engineering Geology. 2022;**297**:106489. DOI: 10.1016/j.enggeo.2021.106489

[60] Guan QZ, Yang ZX, Guo N, Hu Z. Finite element geotechnical analysis incorporating deep learning-based soil model. Computers and Geotechnics. 2023;**154**:105120. DOI: 10.1016/j.compgeo.2022.105120

[61] Xie J, Huang J, Zeng C, Jiang SH, Podlich N. Systematic literature review on data-driven models for predictive maintenance of railway track: Implications in geotechnical engineering. Geosciences. 2020;**10**(11):425. DOI: 10.3390/geosciences10110425

[62] Pei T. Integrating Geotechnical Domain Knowledge into Machine Learning for Slope Stability Predictions. [PhD Thesis]. USA: Penn State University; 2023

[63] Mali N, Dutt V, Uday KV. Determining the geotechnical slope failure factors via ensemble and individual machine learning techniques: A case study in Mandi, India. Frontiers in Earth Science. 2021;**9**. DOI: 10.3389/feart.2021.701837

[64] Xie J, Huang J, Zeng C, Huang S, Burton GJ. A generic framework for geotechnical subsurface modeling with

machine learning. Journal of Rock Mechanics and Geotechnical Engineering. 2022;**14**(5):1366-1379. DOI: 10.1016/j.jrmge.2022.08.001

[65] Zhang W, Pradhan B, Stuyts B, Xu C. Application of artificial intelligence in geotechnical and geohazard investigations. Geological Journal. 2023;**58**(6):2187-2194. DOI: 10.1002/gj.4779

[66] Phoon KK, Zhang LM, Cao ZJ. Special issue on "machine learning and AI in geotechnics". Georisk: Assessment and Management of Risk for Engineered Systems and Geohazards. 2023;**17**(1):1-6. DOI: 10.1080/17499518.2023.2185938

Chapter 5

Application of Machine Learning and Data Mining in Medicine: Opportunities and Considerations

Luwei Li

Abstract

With the continuous development of information technology, machine learning and data mining have gradually found widespread applications across various industries. These technologies delve deeper into uncovering intrinsic patterns through the application of computer science. This trend is especially evident in today's era of advanced artificial intelligence, which marks the anticipated third industrial revolution. By harnessing cutting-edge techniques such as multimodal large-scale models, artificial intelligence is profoundly impacting traditional scientific research methods. The use of machine learning and data mining techniques in medical research has a long-standing history. In addition to traditional methods such as logistic regression, decision trees, and Bayesian analysis, newer technologies such as neural networks, random forests, support vector machines, Histogram-based Gradient Boosting, XGBoost, LightGBM, and CatBoost have gradually gained widespread adoption. Each of these techniques has its own advantages and disadvantages, requiring careful selection based on the specific research objectives in clinical practice. Today, with the emergence of large language models such as ChatGPT 3.5, machine learning and data mining are gaining new meanings and application prospects. ChatGPT offers benefits such as optimized code algorithms and ease of use, saving time and enhancing efficiency for medical researchers. It is worth promoting the use of ChatGPT in clinical research.

Keywords: machine learning, data mining, medicine, artificial intelligence, interaction

1. Introduction

Data mining techniques [1] are computer algorithms used to identify associations from vast amounts of data. They employ heuristic methods, which essentially involve searching for relevant features and patterns and then extracting more feature information to support better decision-making [2]. Based on the algorithms used, data mining can be categorized into several types [3], with the most commonly used including machine learning algorithms, decision tree algorithms, association rule algorithms, and neural network algorithms.

Machine learning algorithms [4] learn from data and often build a model for prediction and classification. They are highly useful in the realm of big data, as they discover patterns from data to provide support for decision-makers. This algorithm can predict outcomes of features in a dataset and can be applied in medical research for tasks such as identifying important factors and their interactions, predicting disease risks, and performing classification tasks based on observed signals in the data, such as disease prediction analysis.

Decision tree algorithms [5] possess strong representational capabilities and are widely applied in data mining. The gradient boosting tree algorithm, based on decision trees, is even more powerful. Its fundamental idea involves inferencing a final decision from a series of feature attributes using scenario analysis. Decision tree algorithms allow the model to create a classification framework based on decision trees in a supervised learning manner, making rational decisions from each node to achieve a high degree of accuracy in prediction tasks.

Association rule algorithms [6] discover relationships among multiple items in large datasets and draw conclusions [7]. Built on association rules and pattern discovery, this algorithm divides data by association degree to uncover patterns. Association algorithms can have HL patterns (high support and low confidence patterns), LH patterns (low support and high confidence patterns), and LL patterns (low support and low confidence patterns). LL patterns may seem insignificant, but they have relevance in specialized fields like medicine. For instance, in rare medical scenarios with high risks, LL patterns can emerge. Association rule's positive patterns also find applications in e-commerce, financial analysis, insurance assessment, and other domains.

Neural network algorithms [8], a branch of traditional artificial intelligence, employ clusters of neurons to create an abstract model for generating outputs from inputs. They can represent data types and build predictive models to respond to complex connections within extensive datasets and make decisions based on these connections. They are commonly used in fields such as clinical diagnostic decision-making and automatic classification.

Artificial intelligence (AI) is a field within computer science [9] that aims to mimic human thought processes, learning abilities, and knowledge storage [10]. In the era of big data, AI technologies can utilize vast clinical datasets to support clinical decisions, unveil latent disease subtypes, associations, prognosis indicators, and generate testable hypotheses. AI is gradually transforming the way doctors make clinical decisions and diagnoses [11]. Machine learning is a crucial branch of artificial intelligence. Machine learning and deep learning techniques have shown superior capabilities in handling large, complex, nonlinear, and multidimensional data compared to traditional statistical methods. They have found widespread applications in the medical field [12]. In the following discussion, I will approach the topic from a medical perspective and provide an overview of the latest applications of AI, specifically using machine learning algorithms such as logistic regression, linear regression, random forest, support vector machines, decision tree algorithms, and neural network algorithms. I will also share insights into the algorithmic modeling process, especially considering recent advancements such as the application of the new AI technology, ChatGPT 3.5. In addition, I will emphasize the validation aspects of post-algorithmic modeling, such as ROC curves, DCA curves, CIC curves, calibration curves, K-fold cross-validation, and even the construction of confusion matrices. Another focal point will be the application of these techniques in the context of the global backdrop of COVID-19 infections, particularly in the realm of public health interaction.

Machine learning algorithms have consistently been a focus and hot topic within data mining [13], especially in the domain of medical big data research. Research aims for a substantial number of study samples, sophisticated computer algorithms, and advanced statistical analysis theory. Data mining techniques can be used for diagnosis, prediction, classification, constructing predictive models, and analyzing risk factors [14]. The goal is to generate more reliable and widely applicable models, leading to practical applications that enhance the speed and accuracy of medical diagnosis [15]. Ultimately, this contributes to the recovery of human diseases and indirectly accelerates research in medical robotics.

Machine learning algorithms encompass logistic regression, COX regression, linear regression, random forest, support vector machines, and NaiveBayes, in addition to newer developments such as KNN, GBDT, Histogram-based Gradient Boosting, XGBoost [16], LightGBM [17], and CatBoost [18]. These advanced algorithms have demonstrated significant improvements over traditional modeling techniques, particularly CatBoost, XGBoost, and LightGBM. These three algorithms can rival any advanced machine learning algorithm worldwide in terms of performance. However, determining the optimal algorithm for model construction requires practical data modeling and validation in real scenarios.

Speaking from my perspective, I have been involved in research related to algorithmic modeling, including projects such as "Construction of Chronic Disease Prediction Models and Applications Based on Data Mining" and "Artificial Intelligence Learning of HUA Susceptible Gene Molecular Typing and Risk Prediction." I possess research experience in using computer algorithms for modeling and validation across tens of thousands of medical big-data cases. My software background extends from SPSS, MedCalc, R software to ChatGPT 3.5, Python, and other statistical software. Statistical theory deepens in tandem with the continuous exploration of statistical operations. I have also published several academic papers on medical computer modeling and validation of chronic diseases, possessing a certain level of applied experience. Here, I will analyze and share experiences in data mining technique modeling and the current clinical applications.

2. Algorithm introduction

2.1 Logistic regression

Logistic regression (LR) is a generalized linear regression model [19] and one of the most widely used algorithms in clinical applications. Besides its ease of implementation in software such as SPSS, its clinical risk analysis based on odds ratios (OR) is easy to comprehend. Moreover, LR's broad applicability is due to its compatibility with numerous real-world problems. LR algorithms [20] are frequently used in clinical data mining, disease autodiagnosis, economic forecasting, and other domains. They can explore risk factors causing diseases and predict the probability of disease occurrence based on these factors. For instance, LR analysis can yield the weights of independent variables, offering insights into which features serve as risk factors for the outcome variable. Utilizing these weights, one can predict the likelihood of an individual falling ill. Many of my research studies also utilize LR as a foundational model, and it indeed offers predictive capabilities. However, if LR is the primary research tool, I recommend considering whether a linear model is suitable, as many clinical issues are not purely linear. In such cases, constructing a nonlinear regression model, which includes curve equations, segmented regression, spline regression, locally weighted regression, and generalized additive models, is advisable. Prior to building predictive models, I suggest plotting scatter plots to visually assess whether the relationship between independent and dependent variables is linear or nonlinear, thereby selecting the appropriate regression modeling method.

2.2 COX regression

COX regression, on the other hand, is a semiparametric regression model based on time and outcome relationships [21], also known as the proportional hazards model or Cox model. It uses survival outcomes and survival time as dependent variables and can simultaneously analyze the influence of numerous feature factors on survival time. It can analyze data with survival time without requiring the estimation of data survival distribution. Because of these advantages, since its inception, survival curve analysis and COX regression models have found extensive application in medical follow-up studies, particularly in clinically relevant research areas closely tied to survival time, such as malignant tumors and cardiovascular diseases. COX regression is currently one of the most widely used multivariate analysis methods in survival analysis.

2.3 Nomogram

Another intuitive machine learning algorithm is the nomogram. Nomograms represent relationships between multiple independent variables in a plane Cartesian coordinate system using a cluster of nonintersecting line segments [22]. While frequently used in meteorology, they have gained widespread application in the medical field in recent years. Nomograms offer visual and convenient ways to present results based on different equations. Hence, nomograms can be used to depict regression results, including LR and COX regression. However, there is a debate on whether creating nomograms for linear regression is necessary, as the calculations for linear regression are straightforward and the advantages of nomograms might not be as pronounced. For LR and COX regression, nomograms conveniently provide insights into disease risk or proportional hazards. Nomograms are constructed based on regression results, representing multiple line segments that facilitate the calculation of risks for different individuals.

Nomograms are widely used in clinical research [23, 24]. After building an LR model, I have found constructing nomograms based on LR to be a logical next step for elaborating results. More specifically, nomograms are constructed through a multifactor regression model, wherein each influence factor's contribution to the outcome variable (the magnitude of regression coefficients) is scored for each value level. These scores are then summed up to obtain a total score. By establishing a function between the total score and the probability of the outcome event, the predicted value for the individual's outcome event can be calculated.

Nomograms are built on the foundation of multifactor regression analysis, integrating multiple predictive indicators and expressing the relationships between variables through scaled line segments drawn on the same plane. They provide a quantifiable and visual representation of results obtained from regression predictive models, which is crucial for guiding clinical research. They hold significant application value in analyzing disease prognosis, especially for diseases such as malignant tumors [25–27].

2.4 Random forest

Random forest (RF) [28] is a classifier that employs multiple decision trees for training and prediction. RF is essentially an ensemble learning system in machine learning [29]. As an emerging and highly flexible machine learning algorithm, RF has a broad range of applications, from healthcare insurance to medical marketing. It can be used to predict disease risks and the susceptibility of patient populations [30]. RF builds upon decision trees (DT), utilizing knowledge learned from the dataset to classify new data. By setting parameters such as the number of trees and branching conditions, multiple DT models are constructed. The final output is determined by the collective decisions of all the decision trees, achieving optimal classification accuracy that surpasses individual decision trees.

The RF algorithm also incorporates the Bagging approach. To explain intuitively, each decision tree is a classifier, resulting in N classification outcomes for an input sample across N trees. RF aggregates all classification votes and designates the class with the highest vote count as the final output. RF boasts several advantages, such as handling missing values, not requiring dimension reduction for high-dimensional data, and introducing randomness to prevent overfitting. Essentially, RF acts as a versatile powerhouse in the field of machine learning, accommodating a wide array of inputs. It excels in estimating inference mappings, and its versatility renders it almost universally applicable. RF is particularly useful. It does not require extensive parameter tuning and validation such as support vector machines (SVM) and yields higher accuracy [31].

I have used over 10,000 clinical cases to construct RF predictive models for chronic diseases such as diabetes, hypertension, and hyperlipidemia. Analyzing the results from the confusion matrix comparison between the RF model and simultaneously constructed SVM and neural network models, there is a significant gap between the weighted average (weighted avg) and macro average (macro avg) values. This discrepancy might be attributed to the fact that macro avg. is the arithmetic mean of indicators for each class, assigning equal weight to each class and disregarding the issue of class sample imbalance. Thus, in cases of imbalanced class distribution, each class's performance receives equal attention. In contrast, weighted avg. is a weighted average, calculated by averaging on each class with the class sample size used as the weight. Consequently, in the case of imbalanced class distribution, the weighted average method places greater emphasis on classes with larger sample sizes while attenuating the performance of classes with smaller sample sizes. However, overall, the RF algorithm, compared with SVM and neural networks, might perform better when dealing with large datasets, high-dimensional datasets, and datasets with complex decision boundaries. It is worthy of broader adoption in clinical research.

2.5 Support vector machine

Support vector machine (SVM) is a type of generalized linear classifier that performs binary classification on data in a supervised learning manner. Its decision boundary is a maximum-margin hyperplane derived from learning samples, and it is a classifier known for its sparsity and robustness [32]. SVM can conduct nonlinear classification using kernel methods, implicitly mapping input content to a high-dimensional feature space. SVM is a binary classification algorithm that supports both linear and nonlinear classifications. Evolving over time, it can now handle multiclass classification and regression problems through extension.

SVM finds a hyperplane for classification on the training dataset and then uses this hyperplane to classify new data based on whether the result is greater or less than 0. For completely linearly separable problems, a hyperplane can be found directly. Even considering noisy data, introducing a soft margin can still lead to a solution. In cases of linear inseparability, data can be lifted to a higher dimension to achieve linear separability, and the issue of the unknown mapping function introduced by dimensionality augmentation can be resolved using kernel techniques. SVM's greatest advantage is its use of kernel techniques to capture nonlinear relationships. In addition, SVM is highly effective in handling small amounts of high-dimensional data, particularly in classification problems. SVM has widespread applications, including medical diagnostics such as disease detection and cancer diagnosis [33]. It can be used in image recognition, video classification, and medical image classification [34] by extracting feature vectors from images and videos and using them as inputs for training models.

Similarly, in building predictive models for chronic diseases, I have also constructed SVM predictive models. The primary parameters for SVM include regularization parameters and kernel parameters. The main parameters of SVM include the regularization parameter, which is also the penalty coefficient of the error term. The choice of kernel function type is crucial. The SVM algorithm can transform nonlinear problems into linear problems by selecting a kernel function. The gamma value, which represents the coefficient of the kernel function, influences the distance between data points in a high-dimensional space. A higher gamma value results in a more complex decision boundary with greater influence from training samples, whereas a lower gamma value leads to a smoother decision boundary. The default setting is "auto," meaning gamma is automatically selected based on the training data. Strengthening training and adjusting parameter settings on the fly are crucial for SVM because it needs to output model prediction results and probabilities based on the input parameters. Based on the SVM modeling research conducted by the author, if parameters are set improperly, the confusion matrix evaluation of the SVM model, particularly precision regarding positive and negative outcomes, might yield larger errors compared with the random forest (RF) and neural network models. Adjusting model parameters might be necessary in such cases. Through repeated training, SVM's model could achieve higher accuracy probabilities than RF or even XGBoost [35]. However, SVM has the disadvantage of being slower in terms of computation speed compared with algorithms such as artificial neural networks (ANN) and RF. SVM's computational complexity is higher, especially for large-scale and high-dimensional datasets, leading to significant computation time and space requirements. In addition, SVM's training process requires multiple iterations, further increasing the computational complexity. Training an SVM model takes a relatively long time, a notable drawback evident in the author's chronic disease modeling. For clinical research data that require repetitive training, this could be a significant limitation.

2.6 Decision tree

Decision tree (DT) is a machine learning method [36]. A prominent feature of DT is its tree-like structure, wherein each internal node represents a judgment on an attribute, each branch signifies the output of a decision result, and each leaf node represents a classification outcome. In general, DT is a decision analysis method that evaluates disease risk and feasibility by constructing a DT based on known probabilities of various scenarios occurring to calculate the probability of the expected net present value being greater than or equal to 0. This decision analysis method uses probability analysis graphically and is termed a decision tree due to its resemblance to the branches of a tree. In machine learning, DT is a predictive model [37] representing the mapping relationship between object attributes and object values. DT uses algorithms such as ID3, C4.5, and C5.0 to generate trees, employing the concept of entropy from information theory.

DT used for classification is called a classification tree and is a commonly used classification method [38, 39]. It is a form of supervised learning, wherein the term "supervised learning" refers to the process of providing a set of samples, each with a set of attributes and a predefined category. Through learning, a classifier is obtained that can correctly categorize newly encountered objects. This type of machine learning is known as supervised learning. On the other hand, the Classification and Regression Tree (CART) is an extremely effective nonparametric method for both classification and regression [40]. It achieves its prediction goals by constructing binary trees. The CART model is widely employed in statistical fields and data mining techniques. It employs an entirely different approach to constructing predictive criteria compared with traditional statistics, making it easy to understand, use, and interpret [41]. Predictive trees generated by the CART model are often more accurate than algebraic predictive criteria constructed using commonly used statistical methods. This advantage becomes more pronounced as the data become more complex and the number of variables increases. In many of my studies, the decision tree (DT) model has been extensively utilized. When constructing a DT model, selecting the right algorithm is of utmost importance, and the CART algorithm is a highly practical choice. Its capabilities in classification and regression tasks are powerful, making it suitable for handling large-scale and high-dimensional data. Its greatest advantage lies in its ability to produce easily understandable graphical results, meeting the needs of most clinical studies. However, it is important not to overlook the disadvantages of the CART algorithm. For example, it can only generate binary trees and cannot handle multiclass problems. The CART algorithm is prone to overfitting and typically requires prepruning to prevent overfitting. Prepruning involves setting a criterion during the tree growth process to stop growth when that criterion is reached, but this can create "horizon limitation," wherein once a node becomes a leaf node, the possibility of favorable branching operations for its successors is cut off. Therefore, pruning parameter settings need to be carefully analyzed based on clinical statistical experience. In addition, DT can indicate interaction effects and can further be used for multiplicative and additive interaction analysis. This is also a focal point and challenge in clinical etiological research in public health, which will be discussed in detail in the final section.

2.7 Artificial neural networks

Artificial neural networks (ANNs) [42] have been a research focus in the field of artificial intelligence since the 1980s. They serve as a specific manifestation of artificial intelligence, thus, this paper places a strong emphasis on them. ANNs abstract the neural network of the human brain from an information processing perspective, creating a simplified model by assembling nodes in different connection patterns to form various networks. In the medical, engineering, and academic fields, ANNs are often referred to directly as neural networks or neural network-like structures. ANNs [43] are a computational model comprised of numerous interconnected nodes (or neurons). Each node represents a specific output function referred to as an activation function. The connection between every two nodes represents a weighted value for the signal passing through that connection, known as a weight. This is equivalent to the memory of an ANN. The network's output varies based on its connection pattern, weight values, and the specific activation function used. The network itself typically approximates a certain algorithm or function found in nature, or it might express a particular logical strategy. Over the past decade, research on ANNs has made substantial progress.

ANNs have successfully addressed many complex real-world problems in fields such as pattern recognition, intelligent robotics, automatic control, prediction estimation, biology, medicine, and economics, displaying remarkable intelligence [44]. ANN is widely applied in the machine learning domain such as constructing disease diagnosis and prediction models in clinical settings as well as applications such as image recognition and speech recognition [45, 46], thereby extending its use to automated disease detection and self-driving cars. ANN is a highly parallel information processing system with strong adaptive learning capabilities [47]. It does not rely on mathematical models of the research object and it demonstrates robustness to changes in system parameters and external disturbances of the controlled object. ANN can handle complex, multi-input, multi-output nonlinear systems. The fundamental problem addressed by ANN is classification. There are various types of ANNs, including back propagation (BP) neural networks, random neural networks, convolutional neural networks (CNN), long short-term memory networks (LSTM), and multilayer perceptrons (MLP) [48]. The choice of ANN algorithm depends on the research objectives and data types, considering the pros and cons of each algorithm.

ANN is a parallel distributed system that utilizes mechanisms distinct from traditional artificial intelligence and information processing techniques. It overcomes the limitations of traditional logic-based AI in dealing with intuition and unstructured information, making its application in artificial intelligence incredibly versatile. Training an ANN requires a significant amount of time and effort. The types of processing units within the network are divided into three categories: input units, output units, and hidden units. The appropriate ANN algorithm must be selected based on the objectives of clinical research. Four common characteristics of ANNs are nonlinearity, nonlocality, nondeterminacy, and nonconvexity. Their common major drawback, however, is the "black box" effect [49], wherein they exhibit similar functionality to the human brain, producing results that are not entirely explainable.

ANN finds extensive application in medical research, primarily due to its advantages [50, 51]. Specifically, ANN possesses self-learning capabilities. For instance, when implementing medical image recognition, inputting numerous diverse image samples and their corresponding recognition outcomes into ANN enables the network to gradually learn to recognize similar images. Self-learning functionality is particularly significant for predictive modeling. For example, convolutional neural networks (CNN) play a crucial role in research related to color ultrasound, imaging, and electrocardiograms [52]. ANN models will also provide economic forecasts, market predictions, benefit projections, and more, making their potential applications extensive. Furthermore, ANN exhibits associative storage functionality. This type of association can be achieved using the feedback network of an ANN. Connections between neurons are assigned relevant weights, and training algorithms adjust these weights iteratively, minimizing prediction errors and enhancing prediction accuracy.

I have used BP neural networks for modeling, with rectified linear units (ReLU) as the activation function for hidden layers and the sigmoid activation function for the output layer. The sigmoid function, being smooth and differentiable, is more precise than linear functions for classification and exhibits better fault tolerance. The differentiability of the sigmoid function allows for its use in gradient descent. Using the sigmoid function in the output layer restricts output values to a smaller range. As a result, the model's expanded connectivity is better than models such as decision trees (DT), whereas the hidden layers can reveal interconnections between independent variables and the dependent variable. Furthermore, ANN excels in rapidly finding optimized solutions. Searching for an optimized solution for a complex problem often requires significant computational effort. Leveraging a feedback-type neural network designed for a specific problem allows computers to use their high-speed processing capabilities to quickly find optimized solutions. ANN can output the importance values of independent variable features. Based on the comprehensive interrelationships among independent variables, features that have the greatest impact on the dependent variable can be clearly identified. Its ability to find optimal solutions is better than that of logistic regression (LR) and DT models. Regarding ANN parameter settings, various transformations can be attempted on the training and validation sets. If using a multilayer perceptron (MLP) neural network, adding a support set is recommended, enhancing the effectiveness of ANN model training and learning. In my experience with MLP neural networks for modeling, I believe that MLP neural networks are more capable of discovering complex factor relationships. MLP neural networks are a forward-feed supervised learning technique that, by setting parameters based on the data type of the dependent variable, can yield more accurate predictive classifications and probabilities, guiding clinical diagnosis and treatment. However, based on my current research experience, both BP neural networks and MLP neural networks have certain drawbacks, such as slow convergence speed, susceptibility to getting stuck in local minima, and inability to reach global optimal solutions. As for other models such as radial basis function neural networks, CNN, and random neural networks, which I have not yet attempted to model, I will refrain from discussing them at this moment.

In addition to the mentioned models, there are other machine learning models currently applied in clinical medical research, including Histogram-based Gradient Boosting, CatBoost, LightGBM, XGBoost, GBM, and GBDT, which are modeling algorithms based on decision trees. CatBoost, LightGBM, XGBoost, and GBM are advanced machine learning algorithms that have been improved and refined based on the GBM algorithm. The GBM algorithm optimizes the loss function towards the direction of steepest gradient through ordered iterations. These algorithms, including GBM and its derivatives such as Histogram-based Gradient Boosting, XGBoost, LightGBM, and CatBoost, are considered focal points in medical research, partly because they are well-suited for the flat data commonly used in medical studies. However, these algorithms are still relatively underutilized in clinical applications due to their complexity, making it challenging for ordinary medical practitioners to comprehend. Moreover, each of these algorithms has its own advantages and drawbacks, and I currently lack relevant experience to provide further insight. It is hopeful that in the future, these advanced algorithms will gradually be applied to clinical research for modeling and validation, exploring more suitable modeling algorithms for medical studies. Many advanced algorithms build upon decision trees as their foundation. Combining my previous machine learning modeling experience, I believe this is largely because decision trees have numerous advantages. For instance, decision trees can be thought of as sets of if-then rules, making them easy to understand and interpret. They require minimal feature engineering, and they do not demand any prior assumptions. Decision trees can handle missing values well and exhibit robustness, especially after implementing methods to prevent overfitting. The development of decision tree construction techniques does not require expensive computational costs. Models can be quickly built even when dealing with large training sets.

2.8 Naive Bayes algorithm

The Naive Bayes algorithm is also widely used in medical research, although I have not yet personally employed it. Naive Bayes is a classification algorithm [53] that is based on the Bayesian theorem and the assumption of feature independence. It classifies sample data sets using probability statistics knowledge. It combines prior and posterior probabilities to avoid using only subjective biases from prior probabilities or suffering from overfitting by using sample information alone. This algorithm is suitable for medium-sized data mining [54]. Naive Bayes is extensively applied in the medical field as well. It can be used for disease diagnosis, modeling patient symptoms and medical test results to predict possible diseases, and assist doctors in making diagnostic decisions. In addition, it is commonly used in medical image classification, although its performance in handling image data is not as strong as other deep learning methods; it can still be applied to simple image classification tasks. By modeling image features, it can automatically classify images into different categories.

3. Evaluate predictive model performance

Evaluating the quality of models involves certain criteria. Regression models are often assessed using metrics such as mean squared error, R-squared, and root mean squared error. In medical research, commonly used metrics for classification models include accuracy, precision, recall, and F1 score, which are derived from a confusion matrix. ROC curve [55] is utilized to derive metrics such as AUC, specificity, sensitivity, and Youden's J index. Furthermore, the clinical impact curve, DCA curve [24], and calibration curve are employed for assessing clinical model performance. These standards are applied in postmodeling validation.

4. Application of artificial intelligence

Building upon the previous explanation, let us delve into the application of artificial intelligence in medical research modeling. Artificial intelligence (AI) is a new technological science that involves the study, development, and application of theory, methods, techniques, and systems for simulating, extending, and expanding human intelligence. AI is a driving force in the new round of technological revolution and industrial transformation. It is an important component of the discipline of intelligence, aiming to understand intelligence and produce intelligent machines capable of reacting in ways similar to human intelligence. This field encompasses robotics, language recognition, image recognition, natural language processing, expert systems, and more.

AI can be implemented on computers in two ways. The first approach employs traditional programming techniques to achieve intelligent behavior without necessarily adhering to methods used by humans or animals. This is known as the engineering approach, which has yielded results in fields such as optical character recognition and computer chess. The second approach, known as the simulation approach, not only focuses on achieving outcomes but also aims for methods similar to those used by humans or biological organisms. Genetic algorithms and ANNs fall into this category.

The recent advancement of AI, such as ChatGPT 3.5 [56], has sparked a new wave of AI learning. I have personally used ChatGPT 3.5 for modeling and validating random forest, support vector machine, and neural network, comparing them with traditional R language-based modeling (RF, SVM, and ANN) through analysis. I have gained some practical experience in this regard. It is important to note that ChatGPT is a language model capable of editing text and computer languages. In the context of medical research, using ChatGPT 3.5 essentially involves coding in computer languages, such as R or Python, to run and obtain results that surpass traditional code-based modeling. Based on test results of ChatGPT 4.0, which achieved over 90% accuracy in various domains such as the U.S. Bar exam, Biology Olympiad, and CPA exam, it is evident that ChatGPT 4.0 possesses enhanced memory, logical analysis, and reasoning capabilities. Its algorithmic models are expected to be more logical and reasoning-oriented. Moreover, ongoing research is exploring the performance of ChatGPT in the U.S. physician exams [57].

ChatGPT boasts distinct advantages, such as enhancing scientific writing, augmenting research fairness and versatility. In the realm of healthcare research [58], its applications encompass effective dataset analysis, code generation, literature reviews, saving time to focus on experimental design, drug discovery, and development, among others. The benefits for healthcare practice involve streamlining workflows, cost-saving, record-keeping, personalized healthcare, and boosting health literacy. In healthcare education, these benefits include enhancing personalized learning and emphasizing critical thinking and problem-based learning [59]. ChatGPT possesses a "brain" reminiscent of humans and can recall previous interactions and user comments, establishing contextual connections—often an area where earlier AI language models lagged behind. Based on these strengths, ChatGPT is gaining traction for extensive use in health care [60], to the extent that it is even listed as a coauthor in many research papers. However, the efficacy [61] and rationale [62] of this practice need evaluation. In addition, developing virtual assistants to aid patients in managing their health is another crucial application of ChatGPT in medicine. ChatGPT can also be utilized for clinical decision support and patient monitoring, suggesting consultations with healthcare professionals based on warning signals and symptoms.

In my research, I have attempted computations using code written by ChatGPT, alongside traditional R language modeling such as RF and ANN. By comprehensively analyzing and comparing the outcomes of modeling using ChatGPT 3.5-generated code for RF, ANN, and others against traditional R language methods, I found that ChatGPT 3.5-based modeling exhibited faster execution, superior accuracy, precision, recall, and simpler parameter settings. The validation performance of models was indeed superior to traditional R language machine learning modeling, making it a valuable approach for clinical promotion and application. Most importantly, models built on AI ChatGPT can automatically adjust model parameters as needed, generating increasingly superior predictive models without the laborious and resource-intensive task of repeatedly training and adjusting parameters, as in traditional R language modeling. AI ChatGPT can continuously optimize algorithmic code during training, unlike traditional R language modeling that lacks real-time algorithm optimization. For healthcare practitioners without coding skills, using AI modeling has tremendous value. Used effectively, ChatGPT can save substantial time for more efficient and prioritized tasks [63]. However, it is important to disclose the shortcomings of using AI ChatGPT in health care and wellness domains [63].

While ChatGPT is currently one of the best editors for computer algorithm code, it does have a time lag and cannot synthesize the latest information to provide optimal algorithmic code. Furthermore, current applications of ChatGPT may raise concerns related to plagiarism, copyright infringement, privacy, and cybersecurity, necessitating a thorough assessment of ChatGPT's security. Lastly, while ChatGPT has certain ethical requirements, it cannot achieve perfection, and it could potentially provide guidance on illegal activities if consulted. Thus, the use of ChatGPT must adhere to relevant laws and regulations and respect regional customs.

5. Introduction to interaction

Moving on to the topic of interaction [64], interactions are classified into multiplicative interaction (product model) and additive interaction (sum model). In the additive model, when there is no interaction, the combined effect of two or more factors acting on an event equals the sum of the effects when these factors act individually. In the multiplicative model, when there is no interaction, the combined effect of two or more factors acting on an event equals the product of the effects when these factors act individually. According to prior epidemiological literature [65], multiplicative interaction indicates statistical interaction effects, whereas additive interaction further suggests biological interaction effects. Given that features in clinical research typically do not have isolated effects, investigating the mutual effects of multiple features on the outcome variable is a direction in clinical research. This aspect becomes even more prominent in public health studies. Similarly, multiplicative and additive interactions do not necessarily coexist. However, in our research, the simultaneous statistical significance of both multiplicative and additive interactions between two features is used as a meaningful criterion for clinical research. In other words, both features must simultaneously fulfill the criteria for statistically significant multiplicative and additive interactions for the evaluation of their combined effect on the outcome variable to have both statistical and biological significance. This is crucial for etiological studies in clinical research. To assess the significance of multiplicative interaction, hypothesis testing criteria can be set, generally at 0.05. However, assessing additive interaction requires the simultaneous fulfillment of three indicators: the relative excess risk due to interaction (RERI), the attributable proportion due to interaction (AP), and the synergy index (S). When there is no additive interaction, the confidence intervals for RERI and AP should include 0, and the confidence interval for S should include 1.

Evaluating the additive and multiplicative models of clinical impact factors holds great significance for disease prediction and diagnosis, particularly in epidemiological statistical research and especially in the aftermath of the COVID-19 pandemic. Calculating additive interaction is important as it has broader implications for public health. I have personally conducted interaction analysis of factors in chronic diseases such as hyperuricemia and hypertension; finally, combining my experience, let us illustrate the importance of interaction research with a simple example.

Suppose a population attends a gathering event, where some individuals have underlying health conditions while others do not. The gathering event is a risk factor for contracting COVID-19, and we want to intervene in order to enhance the control of COVID-19. To optimize the use of resources such as manpower, communication, incentives, and penalties to achieve the maximum impact, we need to consider both the multiplicative and additive interaction models. Assumingly, we apply interaction theory to investigate the combined effects of having an underlying health condition and attending a gathering event on the risk of COVID-19 infection. Let us say that the multiplicative interaction algorithm in this study indicates a reverse multiplicative interaction, implying that among individuals without underlying health conditions, the risk of infection is higher when attending the gathering event compared to not attending. However, relying solely on the results of the multiplicative interaction overlooks the group of individuals with underlying health conditions and may identify the wrong high-risk group. Now, assuming we use the additive interaction algorithm, and it indicates a positive additive interaction, explaining that among individuals with underlying health conditions, intervening in the gathering event can yield greater public health benefits. Therefore, exploring factor interactions can guide clinical and preventive decisions, allowing for the efficient allocation of resources to achieve optimal outcomes. It is important to analyze both the multiplicative and additive models comprehensively to obtain the most reliable research results. With the widespread occurrence of infectious diseases, there is bound to be an increasing amount of statistical research related to factor interactions. Hopefully, this research can provide valuable insights for future infectious disease prevention and control efforts. This serves as one of the most significant contributions of this paper.

Author details

Luwei Li
The First People's Hospital of Nanning, China

*Address all correspondence to: llw135318@gmail.com

References

[1] Yoo I, Alafaireet P, Marinov M, et al. Data mining in healthcare and biomedicine: A survey of the literature. Journal of Medical Systems. 2012;**36**(4):2431-2448

[2] Iavindrasana J, Cohen G, Depeursinge A, et al. Clinical data mining: A review. In: Yearbook of Medical Informatics. US: International Map Industry Association (IMIA); 2009. pp. 121-133

[3] Wu WT, Li YJ, Feng AZ, et al. Data mining in clinical big data: The frequently used databases, steps, and methodological models. Military Medical Research. 2021;**8**(1):44

[4] Deo RC. Machine learning in medicine. Circulation. 2015;**132**(20):1920-1930

[5] Hammann F, Drewe J. Decision tree models for data mining in hit discovery. Expert Opinion on Drug Discovery. 2012;7(4):341-352

[6] Tian H. Brand marketing leveraging the advantage of emoji pack relying on association rule algorithm in data mining technology. Computational Intelligence and Neuroscience. 2022;**2022**:3511211

[7] Hadavi S, Oliaei S, Saidi S, et al. Using data mining and association rules for early diagnosis of Esophageal cancer. The Gulf Journal of Oncology. 2022;**1**(40):38-46

[8] Kriegeskorte N, Golan T. Neural network models and deep learning. Current Biology. 2019;**29**(7):R231-R236

[9] Holmes JH, Sacchi L, Bellazzi R, et al. Artificial intelligence in medicine AIME 2015. Artificial Intelligence in Medicine. 2017;**81**:1-2

[10] Mintz Y, Brodie R. Introduction to artificial intelligence in medicine. Minimally Invasive Therapy & Allied Technologies. 2019;**28**(2):73-81

[11] Hamet P, Tremblay J. Artificial intelligence in medicine. Metabolism. 2017;**69S**:S36-S40

[12] Mentis AA, Garcia I, Jiménez J, Paparoupa M, Xirogianni A, Papandreou A, et al. Artificial intelligence in differential diagnostics of meningitis: A nationwide study. Diagnostics (Basel). 28 Mar 2021;**11**(4):602

[13] Zia A, Aziz M, Popa I, Khan SA, Hamedani AF, Asif AR. Artificial intelligence-based medical data mining. Journal of Personalized Medicine. 24 Aug 2022;**12**(9):1359

[14] Birjandi SM, Khasteh SH. A survey on data mining techniques used in medicine. Journal of Diabetes and Metabolic Disorders. 2021;**20**(2):2055-2071

[15] Wen X, Leng P, Wang J, et al. Clinlabomics: Leveraging clinical laboratory data by data mining strategies. BMC Bioinformatics. 2022;**23**(1):387

[16] Hou N, Li M, He L, et al. Predicting 30-days mortality for MIMIC-III patients with sepsis-3: A machine learning approach using XGboost. Journal of Translational Medicine. 2020;**18**(1):462

[17] Zhu J, Su Y, Liu Z, et al. Real-time biomechanical modelling of the liver using LightGBM model. The International Journal of Medical Robotics. 2022;**18**(6):e2433

[18] Hancock JT, Khoshgoftaar TM. CatBoost for big data: An

interdisciplinary review. Journal of Big Data. 2020;**7**(1):94

[19] Stoltzfus JC. Logistic regression: A brief primer. Academic Emergency Medicine. 2011;**18**(10):1099-1104

[20] Schober P, Vetter TR. Logistic regression in medical research. Anesthesia and Analgesia. 2021;**132**(2): 365-366

[21] Zhang Z, Reinikainen J, Adeleke KA, et al. Time-varying covariates and coefficients in cox regression models. Annals of Translational Medicine. 2018;**6**(7):121

[22] Park SY. Nomogram: An analogue tool to deliver digital knowledge. The Journal of Thoracic and Cardiovascular Surgery. 2018;**155**(4):1793

[23] Wang X, Lu J, Song Z, et al. From past to future: Bibliometric analysis of global research productivity on nomogram (2000-2021). Frontiers in Public Health. 2022;**10**:997713

[24] Zhang W, Ji L, Wang X, et al. Nomogram predicts risk and prognostic factors for bone metastasis of pancreatic cancer: A population-based analysis. Frontiers in Endocrinology (Lausanne). 2021;**12**:752176

[25] Hu C, Yang J, Huang Z, et al. Diagnostic and prognostic nomograms for bone metastasis in hepatocellular carcinoma. BMC Cancer. 2020;**20**(1):494

[26] Yu P, Wu X, Li J, et al. Extrathyroidal extension prediction of papillary thyroid cancer with computed tomography based radiomics nomogram: A Multicenter study. Frontiers in Endocrinology (Lausanne). 2022;**13**:874396

[27] Zhang D, Hu J, Liu Z, et al. Prognostic nomogram in patients with epithelioid sarcoma: A SEER-based study. Cancer Medicine. 2023;**12**(3):3079-3088

[28] Rigatti SJ. Random Forest. Journal of Insurance Medicine. 2017;**47**(1):31-39

[29] Doupe P, Faghmous J, Basu S. Machine learning for health services researchers. Value in Health. 2019;**22**(7):808-815

[30] Guo L, Wang Z, Du Y, et al. Random-forest algorithm based biomarkers in predicting prognosis in the patients with hepatocellular carcinoma. Cancer Cell International. 2020;**20**:251

[31] Uddin S, Khan A, Hossain ME, et al. Comparing different supervised machine learning algorithms for disease prediction. BMC Medical Informatics and Decision Making. 2019;**19**(1):281

[32] Lee EJ, Kim YH, Kim N, et al. Deep into the brain: Artificial intelligence in stroke imaging. Journal of Stroke. 2017;**19**(3):277-285

[33] Gaonkar B, Davatzikos C. Analytic estimation of statistical significance maps for support vector machine based multi-variate image analysis and classification. NeuroImage. 2013;**78**:270-283

[34] Habehh H, Gohel S. Machine learning in healthcare. Current Genomics. 2021;**22**(4):291-300

[35] Silva G, Fagundes TP, Teixeira BC, et al. Machine learning for hypertension prediction: A systematic review. Current Hypertension Reports. 2022;**24**(11):523-533

[36] Al FL, Shomo MI, Alazzam MB, et al. Processing decision tree data using

internet of things (IoT) and artificial intelligence technologies with special reference to medical application. BioMed Research International. 2022;**2022**:8626234

[37] DeGregory KW, Kuiper P, DeSilvio T, et al. A review of machine learning in obesity. Obesity Reviews. 2018;**19**(5):668-685

[38] Zhu Y, Fang J. Logistic regression-based Trichotomous classification tree and its application in medical diagnosis. Medical Decision Making. 2016;**36**(8):973-989

[39] Tsien CL, Fraser HS, Long WJ, et al. Using classification tree and logistic regression methods to diagnose myocardial infarction. Studies in Health Technology and Informatics. 1998;**52** (Pt 1):493-497

[40] Schilling C, Mortimer D, Dalziel K, et al. Using classification and regression trees (CART) to identify prescribing thresholds for cardiovascular disease. PharmacoEconomics. 2016;**34**(2): 195-205

[41] Henrard S, Speybroeck N, Hermans C. Classification and regression tree analysis vs. multivariable linear and logistic regression methods as statistical tools for studying haemophilia. Haemophilia. 2015;**21**(6):715-722

[42] Renganathan V. Overview of artificial neural network models in the biomedical domain. Bratislavské Lekárske Listy. 2019;**120**(7):536-540

[43] Harada T. (2)neural network. No Shinkei Geka. 2020;**48**(2):173-188

[44] Clark JW. Neural network modelling. Physics in Medicine and Biology. 1991;**36**(10):1259-1317

[45] Currie G, Hawk KE, Rohren E, et al. Machine learning and deep learning in medical imaging: Intelligent imaging. Journal of Medical Imaging and Radiation Sciences. 2019;**50**(4):477-487

[46] Ha J, Kim S, Baik Y, et al. Artificial neural network enabling clinically meaningful biological image data generation. In: 2020 42nd Annual International Conference of the IEEE Engineering in Medicine and Biology Society (EMBC) in Conjunction with the 43rd Annual Conference of the Canadian Medical and Biological Engineering Society. Vol. 2020. 2020. pp. 2404-2407

[47] Labdai S, Bounar N, Boulkroune A, et al. Artificial neural network-based adaptive control for a DFIG-based WECS. ISA Transactions. 2022;**128**(Pt B):171-180

[48] Zhang Y, Lin H, Yang Z, et al. Neural network-based approaches for biomedical relation classification: A review. Journal of Biomedical Informatics. 2019;**99**:103294

[49] Nair TM. Building and interpreting artificial neural network models for biological systems. Methods in Molecular Biology. 2021;**2190**:185-194

[50] Khan ZH, Mohapatra SK, Khodiar PK, et al. Artificial neural network and medicine. Indian Journal of Physiology and Pharmacology. 1998;**42**(3):321-342

[51] Cao B, Zhang KC, Wei B, et al. Status quo and future prospects of artificial neural network from the perspective of gastroenterologists. World Journal of Gastroenterology. 2021;**27**(21):2681-2709

[52] Gharehbaghi A, Babic A. Deep time growing neural network vs convolutional neural network for

intelligent phonocardiography. Studies in Health Technology and Informatics. 2022;**295**:491-494

[53] Zhang Z. Naive Bayes classification in R. Annals of Translational Medicine. 2016;**4**(12):241

[54] Cao X, Xing L, Majd E, et al. A systematic evaluation of supervised machine learning algorithms for cell phenotype classification using single-cell RNA sequencing data. Frontiers in Genetics. 2022;**13**:836798

[55] Martinez PJ, Perez MP. ROC curve. Semergen. 2023;**49**(1):101821

[56] Gordijn B, Have HT. ChatGPT: Evolution or revolution? Medicine, Health Care, and Philosophy. 2023;**26**(1):1-2

[57] Gilson A, Safranek CW, Huang T, et al. How does ChatGPT perform on the United States medical licensing examination? The implications of large language models for medical education and knowledge assessment. Journal of Medical Internet Research. 2023;**9**:e45312

[58] Will ChatGPT transform healthcare? Nature Medicine. 2023;**29**(3):505-506

[59] Sallam M. ChatGPT utility in healthcare education, research, and practice: Systematic review on the promising perspectives and valid concerns. Healthcare (Basel). 19 Mar 2023;**11**(6):887

[60] Cascella M, Montomoli J, Bellini V, et al. Evaluating the feasibility of ChatGPT in healthcare: An analysis of multiple clinical and research scenarios. Journal of Medical Systems. 2023;**47**(1):33

[61] Teixeira DSJ. Is ChatGPT a valid author? Nurse Education in Practice. 2023;**68**:103600

[62] Krugel S, Ostermaier A, Uhl M. ChatGPT's inconsistent moral advice influences users' judgment. Scientific Reports. 2023;**13**(1):4569

[63] Dave T, Athaluri SA, Singh S. ChatGPT in medicine: An overview of its applications, advantages, limitations, future prospects, and ethical considerations. Frontiers in Artificial Intelligence. 2023;**6**:1169595

[64] Bohnke JR. Explanation in causal inference: Methods for mediation and interaction. The Quarterly Journal of Experimental Psychology: QJEP (Hove). 2016;**69**(6):1243-1244

[65] Rothman KJ. Epidemiology: An Introduction. New York: Oxford University Press; 2002. pp. 168-180

Chapter 6

Automatic BI-RADS Classification of Breast Magnetic Resonance Medical Records Using Transformer-Based Models for Brazilian Portuguese

Ricardo de Oliveira, Bruno Menezes, Júnia Ortiz and Erick Nascimento

Abstract

This chapter aims to present a classification model for categorizing textual clinical records of breast magnetic resonance imaging, based on lexical, syntactic and semantic analysis of clinical reports according to the Breast Imaging-Reporting and Data System (BI-RADS) classification, using Deep Learning and Natural Language Processing (NLP). The model was developed from transfer learning based on the pre-trained BERTimbau model, BERT model (Bidirectional Encoder Representations from Transformers) trained in Brazilian Portuguese. The dataset is composed of medical reports in Brazilian Portuguese classified into six categories: Inconclusive; Normal or Negative; Certainly Benign Findings; Probably Benign Findings; Suspicious Findings; High Risk of Cancer; Previously Known Malignant Injury. The following models were implemented and compared: Random Forest, SVM, Naïve Bayes, BERTimbau with and without finetuning. The BERTimbau model presented better results, with better performance after finetuning.

Keywords: BI-RADS, deep learning, transformers, BERTimbau, Portuguese NLP

1. Introduction

The healthcare sector is characterized as a large data catalyst, contained in medical records, reports, test results and so on. In the case of textual medical records, the correctly classification of unstructured parts of the texts incorporated into medical documents can support healthcare professionals for managing relevant data effectively and efficiently, organizing the data related to patients and their findings in diagnostic tests.

This work presents a classification system for categorization of BI-RADS [1], based on lexical, syntactic, and semantic analysis of documents, derived from textual

IntechOpen

clinical records, using Deep Learning and NLP (Natural Language Processing). The main goal is to verify the performance of BERTimbau model [2] in BI-RADS categories classification from breast magnetic resonance imaging clinical records. Machine learning models were also used to classify BI-RADS, in order to establish a baseline. The following models were implemented and compared: Random Forest, SVM, Naïve Bayes, BERTimbau with and without finetuning.

After submitting a dataset containing 8813 records of medical texts to deep learning training, a new expert model based on existing rules was created for automated BI-RADS category classification in breast MRI reports, using a supervised machine learning approach. In addition to being able to classify medical texts related to breast MRIs to their corresponding BI-RADS, the model will be able to inform about the quality of the medical record, in relation to pre-existing statistics.

2. Materials and method

2.1 BI-RADS

BI-RADS [1], is an acronym for Breast Imaging-Reporting and Data System, a quality assurance tool originally designed for using in mammography. The system is a collaborative effort of many health care groups but is published and copyrighted by the American College of Radiology (ACR).[1] The system was designed to standardize clinical reporting and is used by medical professionals to communicate a patient's risk of developing breast cancer, particularly for patients with dense breast tissue. The document focuses on patient reports used by medical professionals. The six classification categories of the American College of Radiology are described below.

2.1.1 BI-RADS category 0 - inconclusive

When the radiologist classifies the result as BI-RADS 0 [1], it means that the examination was considered inconclusive or incomplete. Causes for a category 0 include technical factors, such as poor image quality, which may be due to improper breast positioning or patient movement during the exam. Category 0 can also be assigned when there is doubt about the existence or not of an injury, requiring another imaging exam to take the test.

2.1.2 BI-RADS category 1: Normal or negative

When the radiologist classifies the result as BI-RADS 1 [1], it means that no alteration was presented. The exam is completely normal. The breasts are symmetrical and do not present masses, architectural distortions or suspicious calcifications. The risk of malignant lesion in an exam classified as category 1 is 0%.

2.1.3 BI-RADS category 2: Certainly benign findings

When the radiologist classifies the result as BI-RADS 2 [1], it means that some alteration was found in the images, but that the characteristics of the lesion allow us to state that it is benign. To be classified as category 2, the physician needs to be

[1] www.acr.org.

confident in stating that the lesion is of benign origin. If the physician is in doubt, the result cannot be classified as BI-RADS 2, but as BI-RADS 3. Therefore, in practice, a BI-RADS 2 result has the same clinical value as a BI-RADS 1. The risk of malignant lesion is 0%.

2.1.4 BI-RADS category 3 - probably benign findings

When the radiologist classifies the result as BI-RADS [1], it means that some alteration was found in the images, which is probably benign, but which is not 100% safe. As much as the doctor is almost sure that the lesion is benign, if he has the slightest doubt, the classification should be category 3. Therefore, a result in category 3 indicates a lesion with very low risk of malignancy, which does not need to be biopsied initially, but which, as a precaution, should be followed closely over the next 2 years. The risk of malignant lesions in BI-RADS 3 is only 2%, that is, 98% of cases are actually benign lesions.

2.1.5 BI-RADS category 4 - suspicious findings

When the radiologist classifies the result as BI-RADS 4 [1], it means that some alteration was found in the images, which may be cancer, but which is not necessarily cancer. All patients with a BI-RADS 4 result should undergo biopsy of the lesion so that the correct diagnosis can be established. Category 4 is usually divided into 3 subcategories according to cancer risk:

- BI-RADS 4A – Lesion with low suspicion of malignancy – 2 to 10% risk of cancer.
- BI-RADS 4B – Lesion with moderate suspicion of malignancy – 11 to 50% risk of cancer.
- BI-RADS 4C – Lesion with high suspicion of malignancy – 51 to 95% risk of cancer.

Regardless of the BI-RADS 4 subcategory, all cases should undergo biopsy. The difference is that in the patient with BI-RADS 4A, the biopsy is expected to confirm a benign lesion, while in the BI-RADS 4C, the biopsy is expected to confirm the diagnosis of cancer.

2.1.6 BI-RADS category 5 - high cancer risk

When the radiologist classifies the result as BI-RADS 5 [1], it means that some alteration was found in the images, which almost certainly is derived from breast cancer. Breast lesions with typical features of cancer include dense, spiculated nodules, pleomorphic calcifications, lesions with skin retraction or distortions of breast architecture, or fine linear calcifications arranged in a segment of the breast. Thus, all category 5 lesions should be biopsied and the risk of malignancy in a BI-RADS 5 classification is greater than 95%.

2.1.7 BI-RADS category 6: Previously known malignant lesion

The BI-RADS 6 classification [1] is only used in patients who already have a diagnosis of breast cancer established and end up undergoing a diagnostic imaging exam to monitor the disease, for example, after the onset of chemotherapy. This classification serves only to confirm to the physician that the malignant lesion identified in the mammogram is the same previously known.

2.2 Dataset

For this study, 8813 instances of reports issued by a radiology service, fully anonymized, for breast MRIs, comprised between April 2016 and December 2021 were collected. For bilateral breast MRIs, 7360 instances, representing 83.51% of the total number of instances; for resonances of left breasts, 750 instances, representing 8.51%; for MRIs of right breasts, 657 instances, representing 7.45%; and for breast MRIs using the mammotomy technique, 46 instances, representing 0.52%.

The medical record with the highest number of words had a value of 484; the smallest, 130. The average number of words found was 202.

Breast MRI scans grade BI-RADS if indicated. With that, due to the standardization existing in the instances, there is a BI-RADS classification at the conclusion of medical reports. To extract this information and the population of a specific variable, the loc method was used, combined with the str.contains function to extract keywords related to the BI-RADS categories contained in the medical records.

For the algorithms to work more efficiently, a new variable was created, containing the mapped information, but numerically. The variable is then represented like this:

BI-RADS by Category (numeric variable (Category_Code)):

- 2, 3387 instances;
- 3, 1115 instances;
- 4, 971 instances;
- 6, 723 instances;
- 1586;
- 5, 352 instances;
- 4A, 280 instances;
- 0, 10 instances;
- 4C, 9 instances.

The data are naturally "unbalanced", as this reflects what is actually found in a population that undergoes this type of technique for diagnosing breast cancer, which is an expected behavior for such a set and according to the classes observed.

Some of the most common indications for magnetic resonance imaging of the breasts are clarification of inconclusive findings on mammography and/or ultrasound, as well as tracking high-risk patients, not being indicated for initial investigation, as are, for example, diagnostic exams via mammography (**Figure 1**).

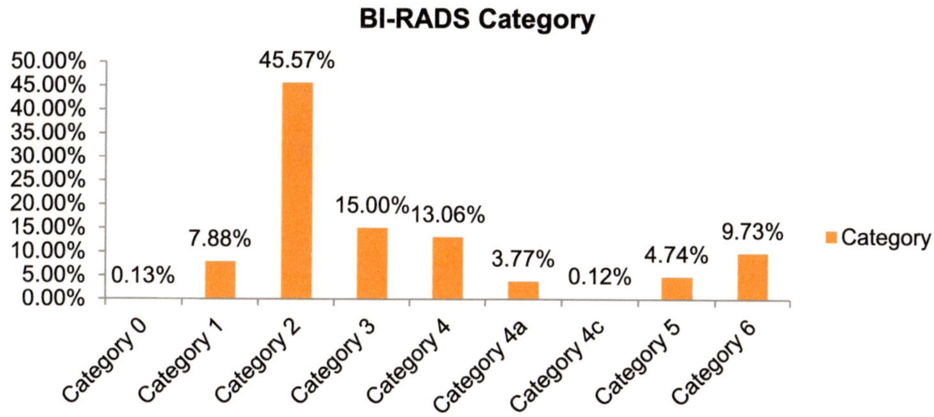

Figure 1.
Percentage of records by class.

2.3 Models

The Transformer architecture (Vaswani et al.) is a natural language processing neural network architecture developed by Google in 2017. It was introduced in the paper "Attention Is All You Need" [3] and has revolutionized the way neural networks are trained to handle language processing tasks such as automatic translation and text generation. The main innovation of the Transformer architecture is the use of attention, which allows the network to consider all input words simultaneously when producing an output. This helps to deal with the variable length dependency problem present in many natural language processing tasks. In addition, the Transformer architecture uses multi-header layers of attention, which helps extend the network's modeling capability.

BERT (Bidirectional Encoder Representations from Transformers) [4] is a language pre-training technique developed by Google in 2018. It uses the Transformer architecture to learn bidirectional representations of each word in a text corpus. This means that, unlike other pre-training techniques that only consider the left or right context of each word, BERT considers the left and right context of each word simultaneously. This allows the model to learn richer and more accurate representations of the words. BERT was trained on a large amount of text from the internet and can be easily adapted to various natural language processing tasks such as text classification, entity extraction and question-answering. It has shown excellent results in many natural language processing tasks and has become a basis for many other language models.

Language model pre-training has been shown to be effective in improving many tasks related to natural language processing [5]. This includes sentence-level tasks such as natural language inference [6], which aim to predict relationships between sentences by analyzing them holistically [7], as well as token-level tasks, such as named entity recognition and answering queries, where models are needed to produce token-level output [8].

In this study, data were submitted to a neural network algorithm called BERTimbau [2], for natural language processing (NLP) in Portuguese, a variation of the BERT algorithm [9]. Random Forest, Support Vector Machine (SVM) and Naïve Bayes machine learning algorithms were also used in order to create a baseline.

Machine learning [10] is a sub-area of artificial intelligence that has shown enormous growth in recent decades. These are mathematical, statistical and computational algorithms that are capable of carrying out an inference process through example-based learning.

Random Forest [11] is a popular machine learning algorithm that belongs to the supervised learning technique. It can be used for classification and regression problems in machine learning. It is based on the concept of ensemble learning, which is a process of combining several classifiers to solve a complex problem and improve model performance.

SVM [12] is one of the most popular supervised learning algorithms used for classification and regression problems. However, it is primarily used for classification problems in machine learning.

The Naïve Bayes algorithm [13] is a supervised learning algorithm based on Bayes' theorem and used to solve classification problems. It is primarily used in classifying text that includes a high-dimensional training dataset. Naïve Bayes Classifier is one of the simplest and most effective classification algorithms that helps in building fast machine learning models that can make fast predictions.

3. Results

3.1 Exploratory analysis

With the dataset still having its original characteristics, in terms of the variable that stores the medical records, the size distribution (number of words per medical record) – see **Figure 2**.

The distribution of the number of existing words per document for each category in the original dataset is presented in **Figure 3**.

In order to clean the data and to improve computational performance, pre-processing techniques were applied such as lowercase, besides of \r, \n, punctuation and stopwords (for Portuguese) removal. The experiments in this work were performed using the dataset in its original characteristic.

Figure 2.
Report size distribution. Original dataset.

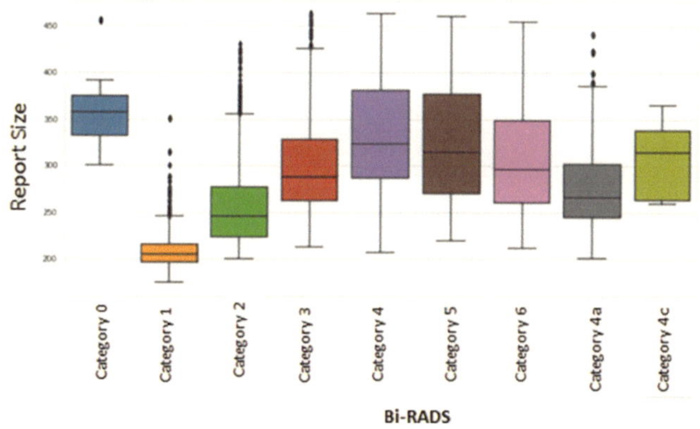

Figure 3.
Boxplot of words in reports. Original dataset.

3.2 Classification models

The original dataset was submitted to three machine learning algorithms (Random Forest, SVM and Naïve Bayes) and a deep learning algorithm (BERTimbau).

The metrics applied to verify the performance of the models were:

Precision: The ability of a classification model to identify only relevant data points. Mathematically, precision is the number of true positives (VP) divided by the number of true positives (VP) plus the number of false positives (FN): VP/(VP + FP); *Recall*: The ability of a model to find all relevant cases in a dataset. Mathematically, recall is defined as the number of true positives (VP) divided by the number of true positives (VP) plus the number of false negatives (FN): VP/(VP + FN); *F1-score*: is defined as the harmonic mean of precision (P) and recall (S). The harmonic mean is an alternative metric to the more common arithmetic mean. It is often useful when calculating an average rate: 2 x (PxS) / (P + S); *Accuracy*: is the number of data points correctly predicted from all data points. More formally, it is defined as the number of true positives (VP) and true negatives (VN) divided by the number of true positives (VP), true negatives (VN), false positives (FP) and false negatives (FN): (VP + VN) / N.

It is important to note that when submitting the dataset to an attribute selection technique, for the machine learning models, categories 0 and 43 were excluded, as their number of instances were inexpressive for the performance of the models. For the Random Forest algorithm, Randomized Search Cross Validation and Grid Search Cross Validation techniques were applied.

The best hyperparameters found with Random Search were:

Bootstrap = False. Method for sampling data points (with or without replacement); max_depth = 30. The max_depth of a tree in Random Forest is defined as the longest path between the root node and the leaf node; max_features = sqrt. This is similar to the maximum number of resources given to each tree in a random forest; min_samples_leaf = 1. Specifies the minimum number of samples that must be present in the leaf node after splitting a node; min_samples_split = 5. Parameter that tells the decision tree in a random forest the minimum number of observations needed at any node to split it; n_estimators = 800. Number of trees in the forest.

Table 1 presents the results found for Random Forest after training.

The confusion matrix, which shows the classification frequencies for each class in the model, for the Random Forest model results, is presented in **Figure 4**.

Random Forest				
Class	**Precision**	**Recall**	**F1-Score**	**Support**
1	0.96	0.92	0.94	84
2	0.92	0.96	0.94	533
3	0,87	0.8	0.83	157
4	0.79	0.82	0.81	145
5	0.59	0,22	0.32	46
6	0.77	0.9	0.83	115
41	0.85	0.83	0.84	35
Accuracy			0.88	1115
Macro AVG	0.82	0.78	0.79	1115
Wighted AVG	0.87	0.88	0.87	1115

Table 1.
Random Forest results.

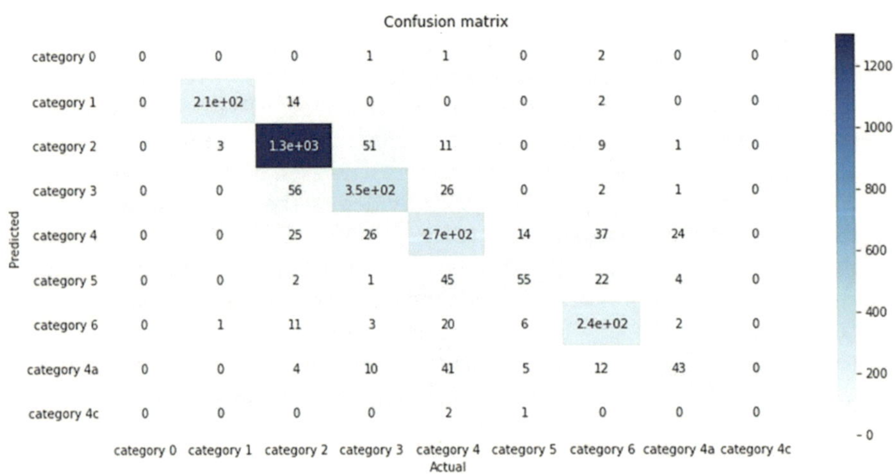

Figure 4.
Random Forest confusion matrix.

For the SVM algorithm, the Randomized Search Cross Validation technique was applied. The best hyperparameters found with Random Search were:

Probability = True, enable probability estimates; Kernel = poly, specifying the kernel type (in this case, polynomial) to be used in the algorithm; Gamma = 10, kernel coefficient for what was specified in hyperparameter Kernel = poly; Degree = 4, Degree of polynomial kernel function (poly); C = 0.01, being the regularization parameter. The strength of the regularization is inversely proportional to C. It must be strictly positive. The penalty is a l2 squared penalty.

Table 2 shows the results found for the SVM after training.

The confusion matrix for the SVM-based model is presented in **Figure 5**.

The values found after training the Naïve Bayes model are presented in **Table 3**. **Figure 6** presents the confusion matrix for Naive Bayes-based model results.

SVM				
Class	**Precision**	**Recall**	**F1-Score**	**Support**
1	1	0.6	0.75	84
2	0.64	0.99	0.78	533
3	0.95	0.39	0.56	157
4	0.56	0.35	0.43	145
5	0.46	0.13	0.2	46
6	0.84	0.37	0.52	115
41	0.67	0.29	0.4	35
Accuracy			0.67	1115
Macro AVG	0.73	0.45	0.52	1115
Wighted AVG	0.71	0.67	0.64	1115

Table 2.
SVM results.

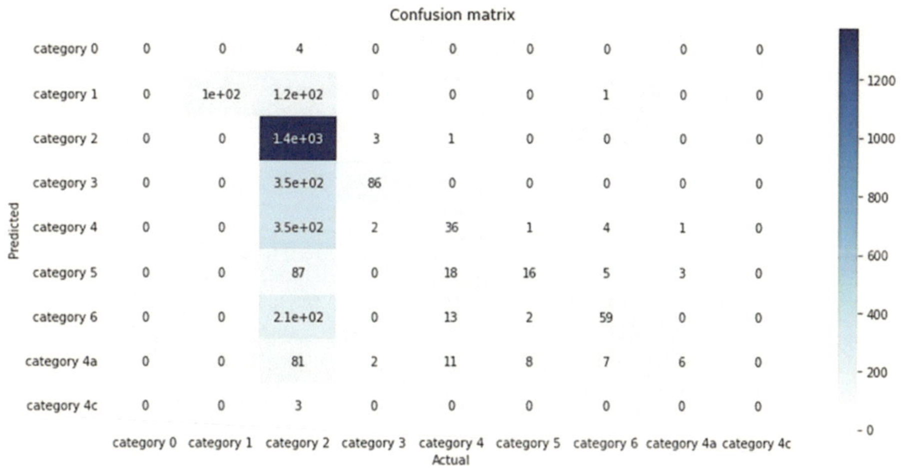

Figure 5.
SVM confusion matrix.

Table 4 presents a summary of the machine learning models results for comparison, which shows that Random Forest algorithm presented the best result.

To submit the dataset to the BERTimbau algorithm, the One-Hot Encoding technique was adopted, transforming the categorical variables into binary ones, with a one-hot encoding being a representation of categorical variables as binary vectors. Specific test steps were applied, with and without finetuning. **Table 5** presents values found after submitting the dataset to four epochs training.

The optimizer used was AdamW with the following parameters: optimizer = AdamW(optimizer_grouped_parameters, lr = 2e-5, correct_bias = True).

The custom optimization parameters were 'params' with the following rule [p for n, p in param_optimizer if not any(nd in n for nd in no_decay)] being the value for no_decay equal to ['bias', 'gamma', 'beta'], which is an iterable thing of parameters to optimize

Naive Bayes				
Class	**Precision**	**Recall**	**F1-Score**	**Support**
1	0.88	0.27	0.42	84
2	0.8	0.98	0.88	533
3	0.81	0.61	0.69	157
4	0.56	0.78	0.65	145
5	0.5	0.04	0.08	46
6	0.75	0.65	0.7	115
41	0.75	0.34	0.47	35
Accuracy			0.75	1115
Macro AVG	0.72	0.52	0.56	1115
Wighted AVG	0.76	0.75	0.72	1115

Table 3.
Naive Bayes results.

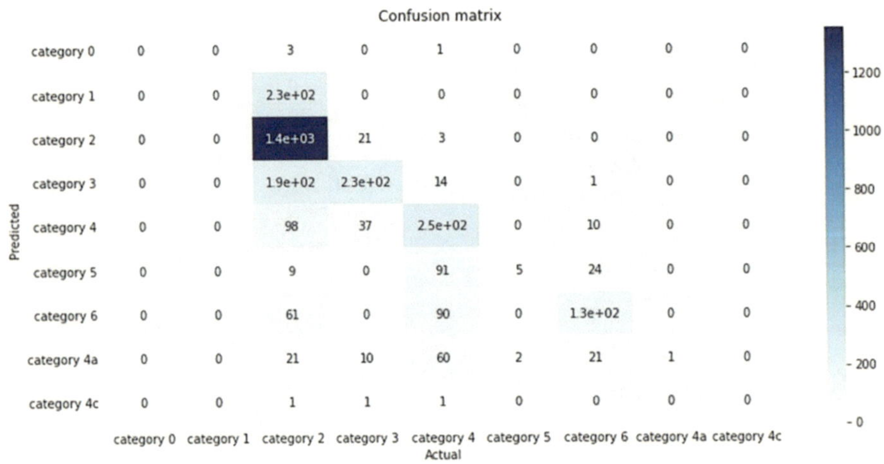

Figure 6.
Naive Bayes confusion matrix.

Test set scores		
Model	**F1**	**Accuracy**
Random Forest	0.787	0.876
Naive Bayes	0.556	0.753
SVM	0.519	0,674

Table 4.
Summary of machine learning models results.

or dictionaries that define groups of parameters; 'weight_decay_rate' with value 0.01 which is the decoupled weight decay to apply or 'params' with the rule [p for n, p in param_optimizer if any(nd in n for nd in no_decay)] and 'weight_decay_rate': 0.0.

BERTimbau				
Class	**Precision**	**Recall**	**F1-Score**	**Support**
birads0	0	0	0	10
birads1	1	0.97	0.98	586
birads2	0.99	0.98	0.99	3387
birads3	0.95	0.97	0.96	1115
birads4	0.86	0.95	0.9	971
birads4a	1	0.7	0.83	280
birads4c	0	0	0	9
birads5	0.51	0.79	0.62	352
birads6	0.93	0.55	0.7	723
Micro AVG	0.93	0.91	0.92	7433
Macro AVG	0.69	0.66	0.66	7433
Weighted AVG	0.94	0.91	0.92	7433
Samples AVG	0.91	0.91	0.91	7433
Test F1 Accuracy	0.92			
Test Flat Accuracy	0.91			

Table 5.
BERTimbau results.

Applying a finetuning, with the aim of enriching the vocabulary of BERTimbau and thus creating both a new specialist model in the area in question and also a specific tokenizer, 1819 new tokens were added. After training in four epochs, the new model was created, expressing a perplexity at a value of 2.17. Perplexity is a measure of how well a probability distribution or probability model predicts a sample. Can be used to compare probability models. A low perplexity indicates that the probability distribution is good at predicting the sample.

The values found using the created expert model, are presented in **Table 6**.

In general, BERTimbau model presented better results compared to machine learning algorithms. **Figure 7** presents the comparative values between BERTimbau model stages.

By observing the values shown in the table above, it is clearly seen that in the vast majority of situations in which the classes were present, the performance of the adjusted model was better than all previously tested models.

4. Conclusions

The Transformer architecture has become the dominant architecture for natural language processing, frequently outperforming models such as convolutional neural networks and recurrent networks in different tasks [14]. Pre-trained models are able to be trained on generic or specialist sets and, consequently, they are easily adapted to tasks with excellent performance. The architecture is particularly conducive to large corpora pre-training, providing accuracy increase in later tasks, such as text classification, language comprehension, and more.

BERTimbau				
Class	**Precision**	**Recall**	**F1-Score**	**Support**
birads0	0	0	0	10
birads1	1	0.98	0.99	586
birads2	1	0.99	0.99	3387
birads3	0.98	0.99	0.98	1115
birads4	0.95	0.98	0.97	971
birads4a	0.95	0.96	0.96	280
birads4c	0	0	0	9
birads5	0.95	0.8	0.87	352
birads6	0.93	0.96	0.95	723
Micro AVG	0.98	0.97	0.98	7433
Macro AVG	0.75	0.74	0.75	7433
Weighted AVG	0.97	0.97	0.97	7433
Samples AVG	0.97	0.97	0.97	7433
Test F1 Accuracy	0.98			
Test Flat Accuracy	0.97			

Table 6.
Fine-tuned BERTimbau results.

	1 - ORIGINAL BERTIMBAU MODEL BERTimbau Tokenizer and Model			
	precision	recall	f1-score	support
birads0	0	0	0	10
birads1	1	0,97	0,98	586
birads2	0,99	0,98	0,99	3387
birads3	0,95	0,97	0,96	1115
birads4	0,86	0,95	0,9	971
birads4a	1	0,7	0,83	280
birads4c	0	0	0	9
birads5	0,51	0,79	0,62	352
birads6	0,93	0,55	0,7	723
micro avg	0,93	0,91	0,92	7433
macro avg	0,69	0,66	0,66	7433
weighted avg	0,94	0,91	0,92	7433
samples avg	0,91	0,91	0,91	7433

	2 - POST FINE TUNING MODEL BIRADS Tokenizer and Model			
	precision	recall	f1-score	support
birads0	0	0	0	10
birads1	1	0,98	0,99	586
birads2	1	0,99	0,99	3387
birads3	0,98	0,99	0,98	1115
birads4	0,95	0,98	0,97	971
birads4a	0,95	0,96	0,96	280
birads4c	0	0	0	9
birads5	0,95	0,8	0,87	352
birads6	0,93	0,96	0,95	723
micro avg	0,98	0,97	0,98	7433
macro avg	0,75	0,74	0,75	7433
weighted avg	0,97	0,97	0,97	7433
samples avg	0,97	0,97	0,97	7433

	FINE TUNING 1 versus 2		
	precision	recall	f1-score
birads0	EQUAL	EQUAL	EQUAL
birads1	EQUAL	BEST	BEST
birads2	BEST	BEST	EQUAL
birads3	BEST	BEST	BEST
birads4	BEST	BEST	BEST
birads4a	WORSE	BEST	BEST
birads4c	EQUAL	EQUAL	EQUAL
birads5	BEST	BEST	BEST
birads6	EQUAL	BEST	BEST
micro avg	BEST	BEST	BEST
macro avg	BEST	BEST	BEST
weighted avg	BEST	BEST	BEST
samples avg	BEST	BEST	BEST

Figure 7.
Comparison between BERTimbau and fine-tuned BERTimbau.

The idea of using machine learning to classify texts in this work with supervised approach is to develop a classification model based on an initial set of labeled texts, using the reached values as baseline for the project.

BERT is undoubtedly a breakthrough in using deep learning for natural language processing. The progress is very significant when it comes to the Portuguese language. The accessibility and fast fine-tuning provide a wide range of practical applications, including using the generated model itself as a basis for creating specialist models in

DOI: http://dx.doi.org/10.5772/intechopen.113886

health area, for example. In the case of this study, fine-tuned BERTimbau managed to capture specific information for a generalist area, increasing its vocabulary and becoming a good model for classifying medical records data, structuring data which is normally unstructured.

Conflict of interest

The authors declare no conflict of interest.

Author details

Ricardo de Oliveira[1], Bruno Menezes[1*], Júnia Ortiz[1] and Erick Nascimento[2]

1 Senai Cimatec, Salvador, Brazil

2 University of Surrey, Surrey, UK

*Address all correspondence to: bruno.menezes@fieb.org.br

References

[1] Castro S, M, Tseytlin E, Medvedeva O, Mitchhell K, Visweswaran S, Bekhuis T, et al. Automated annotation and classification of BI-RADS assessment from radiology reports. Journal of Biomedical Informatics. 2017

[2] Souza F, Nogueira R, Lotufo R. BERTimbau: Pretrained BERT models for Brazilian Portuguese. In: 9th Brazilian Conference on Intelligent Systems. Rio Grande do Sul, Brazil, October 20-23 (to appear). [S.l.: s.n.]: BRACIS; 2020

[3] Vaswani A, Shazeer N, Parmar N, Uszkoreit J, Jones L, Gomez AN, Kaiser L, Polosukhin I. Available from: https://proceedings.neurips.cc/paper/2017/file/3f5ee243547dee91fbd053c1c4a845aa-Paper.pdf

[4] Devlin J, Chang MW, Lee K, Toutanova K. BERT: Pre-Training of Deep Bidirectional Transformers for Language Understanding. 2019. Available from: https://arxiv.org/abs/1810.04805

[5] Dai Andrew M, Le Quoc V. Semi-Supervised Sequence Learning. Available from: https://arxiv.org/abs/1511.01432. 2015

[6] Bowman SR, Angeli G, Potts C, Manning CD. A Large Annotated Corpus for Learning Natural Language Inference. Available from: https://arxiv.org/abs/1508.05326

[7] Dolan W, B, Brocket C. Automatically constructing a corpus of sentential paraphrases. In: Proceedings of the Third International Workshop on Paraphrasing (IWP2005). 2005

[8] Tjong EF, Tjong S, Sang M, De F. Introduction to the CoNLL-2003 Shared Task: Language-Independent Named Entity Recognition. Disponível em: https://arxiv.org/abs/cs/0306050

[9] Deep Learning Book. disponível em: https://www.deeplearningbook.com.br/o-que-e-bert-bidirectional-encoder-representations-from-transformers/

[10] Rudin C, Wagstaff KL. Machine learning for science and society. Machine Learning. 2014

[11] Tin Kam HO. Random Decision Forests (PDF). In: Proceedings of the 3rd International 9 Conference on Document Analysis and Recognition; Montreal, QC; 14-16 August 1995. 1995. pp. 278-282. Archived from the original (PDF) on 17 April 2016. Retrieved 5 June 2016

[12] Cortes C, Vapnik V. Support-vector networks. Machine Learning. 1995;**20**(3):273-297. DOI: 10.1007/BF00994018

[13] Andrew Mccallum. Graphical Models, Lecture2: Bayesian Network Representation (PDF). [Accessed: 22 October 2019]

[14] Rothman D. Transformers for Natural Language Processing: Build Innovative Deep Neural Network Architectures for NLP with Python, PyTorch, TensorFlow, BERT, RoBERTa, and More. Packt Publishing Ltd; 2021